Palmers
Citrus Handbook
EION SCARROW

Fully updated and revised guide to growing citrus in New Zealand

David Bateman

Acknowledgements

Many thanks to Dr Alistair Curry, Kerikeri Research for his help and advice on many aspects of this book; to Grant Tennant of Copperfield Nurseries, Tauranga for help with the varieties table on pages 40–47; and Helen Carr of Flying Dragon Citrus, Kerikeri.

I acknowledge the University of Florida, Institute of Food and Agricultural Sciences (UF/IFAS) website for information on the dwarfing characteristics of trifoliate orange rootstocks.

I also wish to thank the wonderful people at Kiwicare Corporation Limited for help with the information on alternative sprays in Chapter 11 — it would have been very difficult to write without their help.

First published in 1992 by David Bateman Ltd,
30 Tarndale Grove, Albany, Auckland, New Zealand

www.batemanpublishing.co.nz

This edition published in 2008; reprinted 2013

ISBN 978-1-86953-684-8

Book design & typesetting: Alice Bell
Cover design: Alice Bell
Diagrams: Nicki Lowe, Vanilla Fish Design
Printed in China by Everbest Printing Co. Ltd

Contents

 # Preface

It's hard to believe that citrus fruits were once only eaten by kings and the wealthy — today they are popular and available to everyone. There is something very special about these tangy, sweet fruits. Well known for their vitamin C content and fibre, citrus fruits are highly regarded as a health food and there are few foods as nutritious and popular. New Zealand also has a natural advantage in that citrus trees can be grown from North Cape to Bluff (I have even seen them grown on Stewart Island), which means that a range of fresh fruit is available all year round.

Apart from the benefits of their fruit, citrus trees, with their glossy, evergreen foliage, fragrant blossoms and brightly coloured fruit, make attractive ornamental shrubs to add colour to gardens during winter.

From my experience in talkback radio gardening sessions, newspaper advice columns and garden centre services, I know that the most commonly asked questions regarding citrus relate to problems with cultivation. In the pages that follow, I will take you through the steps required to grow citrus to perfection.

Eion Scarrow
135 Hampton Downs Road
RD2, Te Kauwhata 3782
plantguru@infogen.net.nz
www.digthis.co.nz

1: Introduction

The origins of citrus

The original 10 species of citrus are natives of the subtropical regions of South-east Asia. These species were taken all over the world and have since given rise to numerous varieties and a wide range of hybrids. The first mention of citrus in writings by a European was in 310 BC when Greek historian Theophrastus wrote of the citron seen in Median and Persia (present-day Iran). He described the Median or Persian apple, as it was then known, as having 'an exquisite odour, [the leaves] are used as protection from moths in clothes . . . '

The term *citrus* was first used by Pliny, the Roman naturalist and writer, and he wrote of the citrus as a medicine, antidote against poison, perfume and moth repellent.

As trade routes developed, citrons, lemons and oranges were seen regularly in the West. By 1300 AD, Muslim expansion brought the sour orange, lemon and lime to Spain, and European Crusaders returned home with both fruit and trees.

Citrus reached the New World in 1493 when Columbus took the sweet orange, sour orange, lemon, citron and lime on his second voyage and planted them on the islands of Haiti. Spanish invaders introduced oranges to Mexico in 1518, and by the 1700s Spanish missionaries were planting seeds in California.

The Satsuma mandarin was introduced to Japan from China, and by 1805 mandarin oranges had arrived in England and were rapidly distributed throughout Malta, Sicily and Italy.

Citrus reached Australia in 1788, when colonists planted oranges, lemons and limes at Port Jackson; and from Australia, they were brought to New Zealand.

Today, the names of many citrus varieties trace the path of their travels. 'Seville' oranges are named after a city in southern Spain, 'Valencia' oranges are named after a region in eastern Spain, while 'Persian' and 'Tahiti' limes also relate to places where they were established or where they originate from.

Development in the West

As the Spanish missionaries and explorers spread Christianity into California from Mexico, they set up small mission stations and planted gardens. Heavily thorned cactus hedges protected these gardens and the missionaries were reluctant to give seeds to settlers because they found that their supplies of orange juice and lemonade were a good way to ensure that people attended church.

By 1841, orange trees were well established and settlers who had planted orchards were making a good profit from the trees. The California Gold Rush brought boom times to the area. The railroads enabled the oranges to be transported to Sacramento then shipped on to the miners' camps. Oranges were also sent by rail to St Louis in the east, arriving in good condition after a month of travel.

By the early 1900s, thousands of people were migrating to California, planning to make their fortune in either gold or citrus. New sweet varieties of orange had been introduced from Brazil and the settlers were finding that some areas, with their different growing conditions, were better for particular varieties than others. For example, growers found that in San Diego's coastal climate and soils they could harvest several crops of lemons in one year. As production of citrus fruit increased, California growers formed co-operatives that were to lead to the establishment of labels like 'Sunkist', which is now recognised worldwide.

One of the most important developments in citrus cultivation took place in the 1940s, when researchers developed dwarf varieties of citrus trees. In the orchards, citrus trees grew to 7 metres in width, occupying more than 40 square metres of ground space. By experimenting with new varieties of citrus on different rootstocks, researchers aimed to produce smaller trees that would be suitable for planting in the home garden. It is these 'dwarf' varieties that are nowadays grown by home gardeners.

Numerous *scion* combinations were tried. A scion is a cutting from a superior variety that is grafted onto the roots of another variety. The cutting, for example, may bear superior fruit, while the rootstock may be more disease resistant, more suitable for growing in a certain type of soil, or will perhaps reduce the size of the mature tree. An incredible range of new and successful dwarf varieties has been developed, enabling citrus to be grown in home gardens and in countries outside their normal climatic range. By using containers, citrus trees can be grown in cooler climates as the trees can be brought inside during the winter.

Citrus in New Zealand

Citrus trees were introduced into New Zealand from Australia in the early stages of European settlement. James Kemp, who arrived as a member of a missionary party at Kerikeri in the Bay of Islands, is recorded as having brought the first orange seeds from Sydney in August 1818.

By 1820, the 'rough lemon', or citronelle, had been introduced to New Zealand by missionaries. It was to spread rapidly from the mission stations to the Maori settlements throughout the Auckland region.

Interest in the cultivation of citrus trees continued to develop and by the 1870s larger quantities of commercial orange and lemon cultivars were being introduced from Sydney. This led to the first commercial plantings of citrus between 1875 and 1880. David Hay of Montpellier Nurseries, Hobson Bay, Auckland, distributed orange, lemon and grapefruit cultivars imported from Australia in the late 1880s. This work was continued by Hayward Wright, who, at the turn of the century, set up his own nursery at Avondale in Auckland and imported a wide range of citrus cultivars and rootstocks.

One of the most important introductions was the citrus rootstock,

Poncirus trifoliata, which Wright brought into the country in 1918. This eventually led to the production of higher quality mandarins and oranges. Up to this time, only grapefruit and lemons had been considered suitable for commercial production in New Zealand.

Because New Zealand is situated at the southern climatic limit for citrus growing, it is the relatively frost-free north-eastern coastal areas of the North Island, including Northland, Auckland, Poverty Bay and the Bay of Plenty, that contain the main commercial citrus orchards. On favourable sites, free from extremes of heat and cold, large numbers of trees can achieve high rates of production. Because orchards are spread over a large area, shelter belt plantings and the local microclimate become very important considerations. It is easier for home gardeners to provide a suitable site for a small number of citrus trees, although local conditions still have to be taken into account.

Citrus grown in New Zealand tend to have better flavour and higher sugar levels than in any other country due to New Zealand growers' better techniques and cultivation methods.

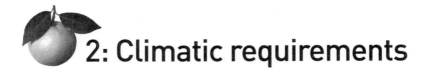

2: Climatic requirements

The effects of climate on citrus

Very few types of citrus being bred in other countries are suitable for the climatic conditions found in New Zealand. Some years ago the Horticulture and Food Research Institute of New Zealand (HortResearch) — or the fruit and trees division of the New Zealand Department of Scientific Research, as it was then called — established small populations of the progeny of crosses between two mandarins, and crosses between various oranges as well as between mandarins and tangelos. This led to an increase in the number of varieties available, suitable for New Zealand conditions.

Gardeners have always liked citrus, and this attraction has often lured them into growing the trees in places where the climate is unsuitable. Citrus generally grow best in places where the climate has mild, moist winters and warm, dry summers. These include the Mediterranean coastline, California, Chile, south-west Africa, parts of Australia and northern New Zealand.

The ancient Romans employed a variety of windbreaks, shelters and other means to enable them to grow citrus trees in the colder, more northerly parts of their empire. Today, citrus trees can be found growing far beyond the boundaries of the world's traditional citrus belt.

The effects of climate on citrus crops can be quite dramatic. In tropical climates, in the zone that stretches 15° north and 15° south of the equator, there are high levels of heat and humidity, rain falls throughout the year, there is little change between day and night temperatures, and dry spells are rare. Under these conditions, citrus fruit remain green on the outside, even when they are ripe on the inside. Citrus crops need cool nights for the skin of the fruit to become brightly coloured.

Citrus trees grow quickly in tropical climates and bloom prolifically after heavy rains. As a result, the trees flower regularly and can often be found bearing fruit at different stages of maturity throughout the year. This makes harvesting difficult, because both the immature and ripe fruit are the same shade of green. Ripe fruit is also prone to ageing rapidly on the tree in tropical areas because of the heat and humidity.

New Zealand covers a wide range of latitudes. The northern parts of the country experience a subtropical climate naturally suitable for commercial citrus production. Northland is the same distance from the equator as California and features warm, humid summers, cool winters and some frosts. But there are other areas in New Zealand where citrus can be grown successfully. Many commercial citrus orchards are in the Waikato and Bay of Plenty and, even in southern New Zealand, where conditions are colder, there are microclimates where citrus can be grown quite well.

New Zealand's climate, with its wide range of daily temperatures, influences the flavour of the fruit, and produces the bright colour we are familiar with in mature fruit. The wide range of temperatures from day to night promotes the formation of both sugar and acid in citrus fruit. Citrus flavour is a factor of both the acid and sugar content. In the tropics the acidic flavour is not as pronounced as it is in subtropical citrus fruits. Tropical citrus are sweeter but subtropical fruit have a better ratio of sugar to acid and, as a result, are considered to be richer in flavour.

High temperatures promote ripeness and a sweet taste in citrus crops. Some citrus varieties require high temperatures that can only be met in the warmest locations. The true grapefruit, *Citrus paradisi*, for example, cannot be grown in New Zealand because high temperatures are required to reduce the bitterness of the fruit at maturity. The grapefruit grown in New Zealand are actually pummelo-orange hybrids.

The length of time grapefruit can remain on a tree when they are approaching maturity is one of their great advantages. In New Zealand's cooler temperatures, grapefruit can be left on the tree for months without deteriorating. This makes them an ideal crop for the home gardener, as they can be left to gradually sweeten on the tree to cater for individual tastes. Citrus crops with a high acid content, such as lemons and limes, have low heat requirements and will ripen without difficulty in cool climates.

Climate also has an effect on a number of other fruit characteristics. Humidity, for example, can have an influence on the size, shape and juice content of citrus fruit. This is why it is important to understand the origins of each type of citrus and its climatic requirements. An understanding of the effects of climate helps a grower decide what steps need to be taken to produce the best results from their particular climate.

Growing citrus in cooler climates

There is an important link between the quality of citrus fruit and the latitude at which it is grown. Areas further south (or north) of the equator produce better quality fruit, though the cooler temperatures present the threat of frost damage. This is important for commercial citrus growers planting large areas; though sometimes growers are prepared to expose their trees to the threat of a severe frost because of the increased returns they receive for better quality fruit.

However, latitude is not critical for home gardeners, as they are able to select sites to suit individual trees and provide shelter if required. New Zealand's cool, mild, climatic conditions provide a number of advantages for home gardeners planting citrus. The fruit is generally of a better flavour and colour and is slow to deteriorate on the tree.

Despite the damaging effects of frosts, citrus trees are surprisingly resilient and can usually stand light frosts (down to -5°C) with relatively minor damage. If frost destroys the leaves on a tree in a home garden, it is often able to recover and produce fruit the following year. There are a number of ways you can reduce the effects of cold.

A good place to start is by selecting varieties that are more cold resistant. Different types of citrus show varying degrees of hardiness. Although precise temperatures at which cold will damage citrus cannot be predicted, the citrus species, from the most tender to the most hardy, are as follows: citron, lime, lemon, grapefruit, pummelo, tangelo, sweet orange, mandarin, Meyer lemon, and kumquat.

In well-drained soils that don't get too wet during frosty weather, citrus can withstand quite heavy frost conditions. Also, if the danger of heavy frost damage can be predicted, trees can be protected with a frost-protection material such as the white, woven frost cloth (available at most garden centres). However, all ripe fruit will be harmed by frost, especially on the side of the tree where the sun strikes first. Young, succulent growth and blossoms are also very tender, making late spring frosts the most damaging.

The use of *Poncirus trifoliata* rootstock, also known as 'trifoliata' or 'tri', will produce a hardier tree. By grafting onto this rootstock the dormancy or inactive period of the scion variety (the young shoot or twig which has been used for the grafting) will be prolonged. This is an acceptable practice in colder climates as it results in increased resistance to frosts. It is also worthwhile selecting early ripening varieties such as Satsuma mandarins as they can be picked before the cold weather arrives.

There are a number of variables that affect cold hardiness:
- The duration of the cold spell is important. For example, a few minutes below freezing is less damaging than an hour below.
- A healthy tree with plenty of foliage cover reduces the risk of cold damage to the fruit.
- What shelter does the tree have? Is it next to a warm wall or is it exposed?
- Is the soil heavy, cold, wet clay? If so, this will damage citrus roots very quickly.
- Is there good 'air drainage'? Colder air moves to low areas in a garden and can damage citrus trees.

Once you have chosen a suitable, healthy citrus variety, the next important consideration is where to locate the tree in your garden.

Location in your garden

Nearly every garden has a range of conditions within it, so take advantage of the microclimates that exist in your garden.
A microclimate is the way the sun, heat and cold all interact with the physical objects in an area. The area may be as large as a valley or as small as a corner of your home garden.

Understanding the microclimate in an area requires little more than simple observation and common sense. Watch the path of the sun and look at the ways its rays reach into your garden during the course of the day and you'll soon work out the warmest growing positions. Because the sun rises in the east and moves northwards to the west, it is the northern side of a house that is likely to produce the warmest and most sheltered positions.

During summer, the sun traces its path through the sky at a higher angle, which means conditions will vary with the seasons. North-facing and south-facing hill sites also have different conditions, while hedges, walls, buildings and other structures will all have an additional effect on the microclimate of an area around your home.

Reflected heat is another factor to consider when deciding on a suitable growing position. Dark-coloured surfaces absorb heat while light-coloured surfaces reflect heat. Stone, cement and masonry will absorb heat during the day, releasing it at night as the surrounding air temperature decreases. Soil absorbs about 30% of the heat that reaches its surface, cement and masonry absorb about 50%, and water absorbs up to 95% of the heat that it receives from sunlight. If you plant a grapefruit tree between a concrete driveway and a north-facing masonry wall, there will probably be enough heat radiated for it to grow superbly, whereas elsewhere in your garden perhaps only a cold-hardy Meyer lemon might survive.

A mulch of grass or leaves generates its own heat as it decays, but this is only minimal; because of the air gaps, only a small amount of solar heat is trapped by the mulch. However, a mulch will help to insulate the root zone (the area radiating out from the trunk where the main feeding roots are situated) when it is spread around a tree. This insulation keeps the soil warmer in winter and cooler in summer. Building up raised beds is also beneficial because the soil drains more quickly and warms up faster.

The wind can also cause heat and moisture loss in citrus trees and can damage the fruit. The effect of wind can be made less severe by using shelter screens or avoided by using the bulk of a house to provide shelter from prevailing winds.

Hot air rises and cold air sinks to the lowest levels, which is why frost damage is more prevalent on a valley floor than on hillsides where cold air is able to 'drain' away. You can minimise the possibility of frost damage in the home garden by not planting in low spots. In the evening, as the ground cools, the cooler air flows down any natural slope. So through clever placement of trees and buildings, you can create good air drainage in your garden, and help to protect your citrus and other tender plants.

Protection against frost

Despite having the right citrus varieties and choosing the correct microclimate, frost still threatens citrus trees in many parts of New Zealand. However, with some protection the trees should be able to survive. Remember, colder climates produce better quality fruit that will not deteriorate so quickly on the tree and which will have a richer flavour. This makes it worthwhile trying to grow citrus in cooler areas.

Frosts occur on cold, clear, winter nights, when the air is still and most of the heat radiation from the sun has been reflected from the ground back up into the atmosphere. Frosts are less likely to occur during cloudy weather as clouds tend to trap solar energy and prevent it from escaping back into space.

Frost damage occurs when ice crystals form inside plant tissue and physically rupture or 'break' the plant's cells. Once the ice crystals have formed on the surface of the plant, they spread rapidly through the tissue. The ice crystals form by a process called nucleation. This process involves minute particles of various types of material, called 'nucleating agents', which act as a catalyst, triggering the growth and formation of tiny ice crystals. Researchers have discovered that there are three main types of bacteria growing on plants that have the ability to act as nucleating agents and bring about the formation of ice crystals.

Experiments conducted to eliminate these ice-nucleating bacteria (INA) have shown that sprays like cupric hydroxide, which is a copper-based chemical, are the most effective. Spraying the trees is a cheaper way to reduce frost damage than conventional methods used by commercial growers, which include fog generators, heaters, wind machines and insulating blankets. At warmer times of the year cupric hydroxide is usually applied at 20-day intervals as a fungus preventative. Care should be taken, however, as this chemical may also cause young leaflets to harden off, which prevents them from growing to full size. For frost protection you need to apply the spray at least seven days prior to any expected frosts, as cupric hydroxide takes several days to kill the INA.

Citrus trees go through regular growth cycles in which masses of new growth are produced, followed by periods of little growth while the new foliage hardens off. It is this new growth that is most susceptible to frost damage. By being careful when you apply fertiliser, you can restrict new growth to times of the year when there is less likelihood of frosts.

Citrus need plenty of nitrogen during their maximum growth period, which is from early spring until December. However, in cooler areas it is important to gradually decrease the amount of nitrogen fertiliser applied as the growing season progresses, ceasing all applications by mid-summer. This will allow the new growth to slow and harden off by autumn, increasing the ability of the tree to withstand cold in winter. Too much nitrogen at the end of summer will produce a mass of new foliage that is more likely to be damaged by frosts (see Chapter 4 on Nutrition).

You can protect the trunks of young trees against frost by wrapping newspapers or some other form of insulating material around them. Small trees can also be protected by building a frame around each tree then covering it with plastic or sacking during cold weather. The cover must be taller and wider than the tree as any foliage that comes into contact with it is liable to be damaged by cold.

Another way to protect crops from frost, especially where spraying may be difficult or you do not wish to spray, is to use frost cloth crop cover. The frost protection properties of this cloth are amazing. It is a non-woven polypropylene weighing only 17 grams per metre and is so light it will not damage even the most wispy of plants. You simply cover your crop or precious trees with it. By creating a microclimate around the plant and raising soil temperatures by up to 4°C it should give you all the protection you need. Light can penetrate through the frost cloth and it can also prevent damage from heavy winds. It comes in large rolls, but most retail outlets sell it by the metre. It lasts for years if you take care of it. Make sure it is thoroughly dry before rolling and storing.

A new technique now also available to both commercial and home gardeners in New Zealand is ThermoMax, a foliar spray, which has been tested by HortResearch. ThermoMax works by increasing the phosphorus metabolism of the plant. This provides an internal warming effect, which is not only useful to protect against frost but also reduces the general effects of cold on fruit set. For more details on the effectiveness of ThermoMax check out www.bdmax.co.nz.

3: Soil preparation

LIKE MOST OTHER subtropical fruit trees, citrus trees will not grow well in poorly drained or waterlogged soils. Their feeding roots are close to the surface so it is also important to keep the ground around the base of the trees free of weeds and other ground-cover plants, which may compete with them for moisture and nutrients. The key words to successful citrus culture are drainage, soil acidity, weed control and irrigation. Most orchardists try to correct drainage deficiencies and any imbalance in the soil acidity level before they plant their trees. This is also a good idea for the home gardener.

Drainage

The importance of good drainage cannot be over-emphasised. In new subdivisions clay soils are usually compacted by the heavy earthmoving machinery used to contour the ground. The soil is compressed to such an extent that it can take years to recover naturally. Many residential gardens have only a shallow topsoil layer on a solid clay base and it is important to break up this clay before planting. If you have any doubt about soil drainage then correct it through one of the two following methods.

You can lay a drainage system such as a strip drain which will carry excess water away, either to a stormwater drain or to a deep sump, which you will have to dig if an existing drain or sump is not available. (See diagrams opposite.) Strip drains are drainage pipes made of non-rotting material covered with a porous synthetic cloth which allows water to percolate through while preventing the drain itself from becoming clogged with soil or clay. The pipes are laid in trenches dug down to a depth of about 1 metre in the area of clay soil you want to drain.

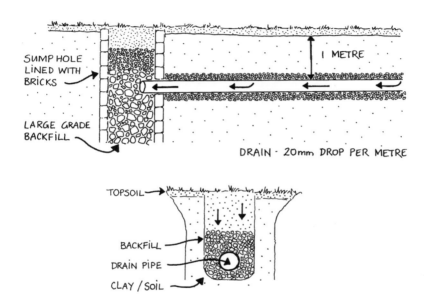

These drawings show the placement of a strip drain in the soil. Water seeps through the soil, along the pipe and drains away down the sump hole.

Another, possibly easier, way to deal with heavy clay soils is to incorporate two large barrow-loads of composty soil (this is soil which comprises at least 30% organic material) into the top layers of the clay soil where you intend to plant your tree. Place the soil in a rounded pile on the clay and lightly fork it into the top 15–20 centimetres. Then plant your citrus into this raised portion. You will need to water the tree carefully over the first summer ensuring that it doesn't get too dry.

The composty soil provides nutrients for the tree and also improves the condition of the clay soil beneath. Earthworms pull the nutritious material from the top layers down into the clay and also create tunnels which aerate the soil, further improving its condition. Deep digging before planting is also important for aerating the soil.

When planting citrus trees on a slope, dig out a hole 1 metre deep, as you would on a flat section. Place a length of strip drain in the base of the hole, emerging out onto the slope, then cover it with a layer of scoria or other suitable material. (See diagram below.) Replace the original soil, mixing in a quantity of compost if it is made up mainly of clay. Sloping sites are much easier to plant citrus trees on than flat sites because they are easier to drain.

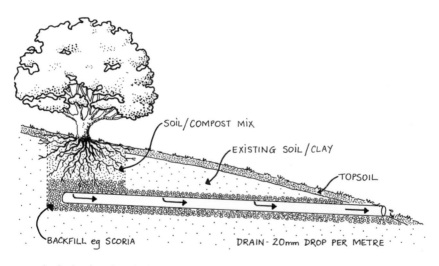

A strip drain placed to drain a tree planted on a sloping site. There is no need for a sump hole as the water can just drain out and down the slope.

Soil acidity

The acidity of the soil has a direct effect on citrus tree growth. It can also influence the uptake of various important nutrients. A soil can be described as acidic, neutral or alkaline. The measurement used to describe the acidity or alkalinity of a soil is the pH scale, which runs from 0 to 14. A measurement of 0 is the most acidic condition. Seven indicates a neutral pH and above 7 the soil is increasingly alkaline. This form of measurement is useful both for home gardeners and commercial growers, as it allows the pH to be measured and then altered to suit different types of trees.

Citrus trees grow well in mildly acidic soils; in the range between 5.8 and 6.5 on the pH scale. Below a pH level of 5.8 and above 7.0 many of the nutrients important for plant growth are less available. *Poncirus trifoliata*, which is the main citrus rootstock in New Zealand, does not grow well in alkaline soil conditions.

There are a number of ways of testing the pH of the soil. Commercial growers usually employ the professional services of a horticultural advisory company, if local knowledge or previous tests are not available. Hydroponic suppliers and many garden centres have home testing kits which you can use to test conditions in the garden. Once the pH level of the soil has been determined, it can then be altered if necessary. You can add lime to raise the pH of acid soils, or compost, sulphate of ammonia or iron chelates to lower the pH of alkaline soils. Routine checking to find out if further corrections are required should be carried out every one or two seasons.

Weed control

When planting citrus into an existing lawn, it is important to keep the grass from returning onto the area dug for the tree. Grass must be kept at least 50 cm outside the 'drip line'. This is the area of soil below the outer spread of the branches, where the feeding roots of the tree are located. Grass will absorb nutrients and moisture in the soil to the detriment of the tree. Grass and weeds may also harbour a number of pests and diseases.

It is important to keep weeds away from the base of the tree as this allows air movement which will keep the skirt of the tree dry. Slugs, snails and a number of diseases are likely to flourish in damp conditions. Phytophthora infection, leading to collar rot, may occur if the trunk is damaged by a motor-mower — another reason why it is not a good idea to allow grass to grow right to the base of the trees.

There are numerous chemical products designed for weed control but take care before using some of them as they can damage the soil by killing the bacteria and fungi living in it. While products such as Network and Roundup are biodegradable and work by entering the leaf systems of plants rather than through the soil, they have now been scientifically shown to damage soils, even though tests done on this family of herbicides have previously suggested they are quite safe to use around citrus trees. It is best to avoid them and use a compost-based mulch comprised of 80% compost and 20% vermicast, which is now readily available in New Zealand.

Vermicast, a by-product of the composting worm, is produced once organic material has passed through the worm's gut. The worm digests its food through the action of the bacteria and enzymes it has in its gut, and large numbers of these are excreted by the worm in the rich black castings called vermicast. These bacteria and enzymes release nutrients in the soil, making them readily available for plants.

Shallow cultivation around the base of citrus trees is another way to control weeds. But be careful not to damage feeding roots close to the surface.

In large citrus orchards, a weed spraying strip is usually maintained under the trees, while clover is allowed to grow between. Clover supports nitrogen-fixing bacteria which grow in nodules on the roots, increasing the quantity of this important element in the soil. When the clover reaches 18 centimetres in height it is cut back to 5 centimetres to minimise the amount of water it takes out of the ground, thereby reducing competition with the citrus trees.

One of the best ways to keep weeds away from the base of a tree is to apply a thick mulch, preferably of compost, but you can use lawn clippings or other organic material. This will help to conserve moisture while also providing a constant supply of additional nutrients into the soil. Worms drag much of the material down into the soil while the lower layers of the mulch break down with the help of insects and decomposing bacteria. The best time to apply a mulch is in spring when the soil is still moist after winter rains. The mulch will help to conserve this moisture and keep the root zone cool throughout the heat of summer. Leave an area of 30 centimetres free around the base of each tree because the build-up of heat as the grass rots could burn the bark. Also, this space allows air movement which will keep the tree trunk dry and free from collar rot.

A variety of organic materials can be used as mulch around trees. Sawdust is suitable but must be from untreated timber and should be applied with blood and bone to prevent nitrogen starvation of the trees, which can occur as soil bacteria break down the sawdust. To each large barrow-load of sawdust add 2 kilograms of blood and bone.

Animal manures, when used annually as a mulch, will produce a gradual build-up of soil acidity. This can be counteracted by applying a dressing of lime in autumn. Apply between 180 grams and 230 grams per square metre. Dolomite lime has the additional benefit of adding magnesium to the soil.

Irrigation

Irrigation is an important factor in producing large crops when citrus trees are grown in shallow soils. Lemons, in particular, need to be well watered to encourage regular growth and several flower flushes each season, as well as bringing fruit up to size. Most commercial growers use trickle irrigation systems featuring microtubes to deliver water throughout drought periods. Home garden irrigation systems are also available but it's best to use them only during dry weather. Excessive irrigation during August and September, when the soil is still cold, can cause damage to young developing roots.

4: Nutrition

THE CORRECT APPLICATION and timing of fertilisers can have a profound effect on the health and cropping capabilities of citrus trees. All new citrus trees should be planted into fertile soil if possible. Mix plenty of compost, peat, leaf mould, stable manure or other organic materials with the soil, especially if working with heavy clay. Do the same with sandy soils to improve moisture retention and prevent excessive leaching of vital nutrients.

Ideally, make sure organic material has plenty of time to break down in a compost heap before mixing it with the soil into which you are going to plant a new tree. This is because the decomposing bacteria that break down organic material generate a considerable amount of heat. The heat build-up in the soil can damage young feeding roots if fresh compost is used. A mulch of fresh compost is also liable to damage the bark if it is applied directly around the trunk. Along with suitable organic material, a 2-kilogram application of blood and bone worked into the soil will give young citrus trees a good start.

Nutrient deficiencies are a common cause of failure of many citrus trees in home gardens. A good all-purpose citrus fertiliser can be made up by combining 2 kilograms of calcium ammonium nitrate or urea, 2.5 kilograms of rock phosphate and 500 grams of muriate of potash. These are available at all good garden centres or can be supplied by stock and station agents. As a general guide, this mixture should be applied at a rate of 500 grams per tree for each year of the tree's age from planting, up to a maximum of 5 kilograms per tree per year. Some gardeners apply one-sixth of the correct yearly amount of fertiliser at two-monthly intervals.

Most garden centres also carry a number of ready-mix citrus fertilisers which are effective and easy to apply.

Animal manures are an excellent source of nutrients for citrus trees. The best time to apply them is early autumn. They contain only very small amounts of nitrogen so don't boost new growth and expose the tree to frost damage. Both manures and all-purpose fertilisers should be applied to citrus trees following a regular nutritional programme. (See the monthly citrus calendar on pages 95–97).

Nutrient deficiencies often prevent citrus trees from achieving their optimum growth and fruit production, while excessive application of nutrient-rich fertilisers and manures can also adversely affect the trees. Too much animal manure can make the soil more acidic, which can reduce the availability of nutrients like molybdenum, magnesium, calcium, sulphur, potassium, phosphorus and nitrogen. Alkaline soils, on the other hand, tend to suppress the availability of copper, zinc, manganese, iron and, again, nitrogen. A soil with a pH range of about 5.8 to 6.5 is optimum for uptake of the greatest range of nutrients.

Commercial growers follow routine programmes of fertiliser application each year but also use soil analysis, leaf analysis and visual analysis of the trees to make sure the correct nutrient balance is achieved. When a large number of trees is involved, this can save fertiliser wastage and ensure optimum fruit production. The different types of analysis are used because each focuses on a different aspect of tree performance. Soil analysis identifies the nutrient balance in the soil. This, however, will not identify problems such as malfunctioning roots, which is why visual analysis is used to try and identify any acute deficiency symptoms. Leaf analysis identifies the nutrient balance within the tree and requires a detailed laboratory assessment of a random sample of leaves from trees in an orchard.

These techniques are too involved for the home gardener, but a basic knowledge of the roles the different nutrient elements play can help you apply fertilisers more confidently.

Nitrogen

Nitrogen is an essential nutrient element for growth, flowering and fruit production. Citrus trees growing in a soil deficient in nitrogen usually look stunted with yellow-coloured foliage, and flower and fruit poorly. However, too much nitrogen will produce an excess of lush, soft growth. The fruit can also be coarse and thick-skinned, take longer to mature and have poor colouring.

Nitrogen-rich fertilisers should be applied just prior to and during flowering, at fruit set and at fruit fall. This will help the tree during the main growth flushes. Most citrus trees have two main growth flushes, when they are using nitrogen rapidly. Lemons usually have three main growth flushes.

The best time to apply nitrogen is at the start of the growth flushes which take place in spring and autumn. Three applications per year are recommended. Because heavy dressings of fertiliser late in the season have such an adverse effect on fruit production, most commercial growers apply at least two-thirds of the annual nitrogen requirement in early spring (September and November), then the remaining third in February. Late applications in autumn can also be dangerous for the tree because they encourage new growth which will not have time to harden off before winter. Like most other trees, citrus store nitrogen in their woody tissues. They can use this stored nitrogen in spring, so symptoms of nitrogen deficiency at this stage are sometimes the result of a shortage in the previous season.

Many citrus trees tend to produce a smaller crop every second year. 'Wheeny' grapefruit, for example, will produce virtually no fruit every second year. Termed 'biennial bearing', this tendency means that as the tree does not require the extra nutrients in the 'off year', applications of nitrogen could be reduced or even eliminated in that year.

You can apply nitrogen to the soil in a number of different forms. Nitrates give the quickest response, followed by urea, then ammonium. Animal manures give the slowest response. Commercial mixes of fertilisers that contain more than 8% available nitrogen are also suitable. You can spray nitrogen-based liquid foliar fertiliser on the leaves of a citrus tree to achieve a quick response. But only small quantities can be applied and they should be used to supplement rather than replace solid fertilisers.

Young trees need frequent applications of nitrogenous fertilisers in small quantities. Never apply fertiliser directly against the base of the trunk as the fertiliser will damage the bark. It is also important to apply fertiliser to moist soil as root burn, followed by leaf scorch and leaf drop, may occur if the soil is dry. This is especially important when nitrogen is being applied as urea.

Now that efficient liquid fertilisers, such as Dig This Organic Liquid Fertiliser, are readily available, it is advisable to use these instead of solid fertilisers. This fertiliser is made from waste from the fishing industry, mainly the leftovers, like bones and guts, which are mixed with seaweed from Norway. Its NPK (nitrogen, phosphorus, potassium) is 8.4.6 and is BioGro certified. See Foliar spraying on page 29 for advice on the application of fertilisers.

Phosphorus

Phosphorus is an important nutrient element, affecting both growth and fruit quality. The skin condition and the pithy parts within the fruit are influenced by phosphorus, which also plays a major role in root growth. When this element is deficient, the trees will exhibit stunting and poor growth. The fruit will have thick skin, low juice content and the juice is often very acidic. The visual symptoms of phosphorus deficiency are not usually seen in citrus trees in New Zealand; however, the effects on fruit quality are common in most districts. Most phosphorus fertilisers should be applied in early spring or autumn, although spring is usually the most convenient time. An excessive application of phosphorus reduces the acid content in the fruit and the concentration of vitamin C.

Potassium

Potassium plays a complex role in plant metabolism. In citrus trees potassium has an important influence on fruit quality, while also affecting the health of the trees and their ability to resist adverse conditions. A deficiency in potassium can cause bronzed and sometimes scorched leaves as well as premature leaf fall. Relatively high rates of potassium improve the quality of lemon fruit, but too much produces large fruit with thick, coarse skins and low juice content. Potassium is usually best applied in a single dressing during spring. It is always found in commercial citrus fertiliser mixes or can be bought separately as sulphate of potash.

Magnesium

Magnesium is an important element in chlorophyll, which plants use to transform minerals into growth-promoting sugars with the help of the sun's energy. The process is called photosynthesis and chlorophyll is the key element, forming the green pigment in the foliage and other parts of the tree. As a result, magnesium deficiency produces a yellowing of foliage with an inverted 'V' remaining green at the base of each leaf.

Premature leaf fall and a tendency towards biennial bearing in seedy citrus varieties are other symptoms of magnesium deficiency. Magnesium plays an important part in seed development and also influences the ability of the tree to take in zinc and manganese. Because magnesium is a relatively mobile nutrient element, deficiency symptoms are usually visible from new shoots to the older foliage, as well as in the developing fruit. Magnesium is usually best applied in spring, but to produce a rapid response when the deficiency is severe you can spray it directly onto the foliage.

Trace elements

Manganese is a trace element and only required in very small quantities. The main deficiency symptom is yellowing between the leaf veins. Initially, this yellowing is blotchy in appearance, with a band of darker green remaining along the midrib and veins of the leaves. Acute deficiency is liable to affect the efficiency of the leaves to the extent that growth of the tree will also be affected.

Although only small amounts of manganese need to be present in the soil, deficiency symptoms can be found in all New Zealand's main citrus districts. Manganese deficiency can sometimes be caused by a high pH level in the soil as this impairs its uptake by the roots. The deficiency is not easily corrected by applications of fertiliser to the soil. Foliar applications usually provide the best solution. Because of this sensitivity to manganese deficiency, it is a good idea to apply serpentine superphosphate, calcined magnesite, dolomite or kieserite to the soil before planting new citrus.

Another trace element that can affect fruit production is zinc. The deficiency symptoms are similar to manganese deficiency, although the yellowing between the leaf veins is less blotchy. The leaves are smaller, narrow, pointed and rosetted. This 'little leaf' symptom is the main distinguishing feature of zinc deficiency. Like manganese deficiency, zinc deficiency is usually corrected by foliar spraying.

Foliar spraying

Foliar spraying is a useful supplement to applications of solid fertilisers in a citrus nutrition programme. Use Dig This Organic Liquid Fertiliser, a liquid blood and bone fertiliser and very efficient foliar spray as it contains all the elements for successful growth (for more information on this fertiliser see digthis.co.nz, or email plantguru@infogen.net.nz). These fertilisers will provide citrus trees with the nutrients that are deficient in New Zealand soils.

Citrus trees also need an application of copper each year. This can be applied as a foliar spray by making up cupric hydroxide or Cuprox at the recommended rate and applying it to the trees. These sprays will control pests as well as provide nutrients. Copper is an essential constituent of a number of enzymes that are involved in many complex life processes in plants. Soils most often found to be copper deficient are acid soils, calcareous soils (soils with a high proportion of chalk or lime), and loams and gravelly soils. Copper deficiency shows up as mottled yellow younger leaves and 'little leaf' syndrome over the whole tree.

If using copper, in any form, as a fungicide to prevent verrucosis (see page 82), this will give plants sufficient copper for their needs.

Foliar sprays are a very effective way to apply nutrients, but care should be taken to make sure not to exceed the label-recommended rates. Also, apply them only on overcast days or in the evening, as bright sunshine on leaves just sprayed with nutrients can result in leaf damage. The higher humidity at these times also aids nutrient uptake. The best time of year to apply foliar sprays is early November when the nutrients can be taken up by the young spring foliage. Young leaves can absorb over half the applied nutrients in less than 24 hours, whereas older leaves take longer and absorb less.

Spray about 200 millilitres onto small trees and 500 millilitres onto larger trees. This amount should be just enough to wet the plants. Repeat the spray after two weeks, and again every month: little and often is better than a large dose.

Remember that while nutrition is important, young trees will not respond to fertiliser applications if they are suffering from poor drainage, lack of shelter, poor soils, root disease or other diseases. (See Chapter 12, page 95, for a full year's nutrition programme.)

5: Propagation techniques

THERE ARE TWO MAIN methods used to propagate citrus trees. They can be grown from cuttings — a simple and effective way of producing a large number of trees. However, it can be difficult to produce a well-balanced tree using this method due to the fact that trees grown from cuttings produce several leaders low down on the trunk, and if these are not pruned out early, a low tree with a clutter of branches will result.

The other propagation method is known as budding onto rootstock. This is the main commercial propagation technique. It involves a bud from the parent citrus tree being grafted onto rootstock grown from seed. This method is also suited to the production of a large number of trees. Many home gardeners prefer to buy these trees already grafted from a nursery.

When propagating citrus trees by budding onto rootstock, the best characteristics of two different varieties can be combined into one tree. The selection of the budwood is critical when trying to produce a tree that will provide regular heavy cropping, while remaining free from pests and diseases. In commercial nurseries the parent trees are given special treatment so they will produce vigorous new spring growth and provide healthy buds. They are fertilised carefully and pruned in early spring. Each shoot usually contains up to 12 buds; however, only the middle buds will be suitable for the propagation of new trees. The selection of the rootstock is also important in determining the hardiness of the tree, its resistance to pests and diseases, adaptability to different soils, its size and the quality of the fruit.

Three types of seeds are used to provide rootstock; they are sweet orange, trifoliata and rough lemon.

Poncirus trifoliata, also known as 'trifoliate orange', 'trifoliata' and 'tri', is the main rootstock used for citrus in New Zealand. It is a deciduous tree and it is this factor that makes the rootstock desirable because growth slows down in the autumn, leaving behind firm, ripened growth. It is also a relatively freeze-hardy rootstock, which makes it particularly suitable for New Zealand conditions. It is the ideal rootstock for oranges, mandarins, tangelos and grapefruit. Ironically, *Poncirus trifoliata* is not a true citrus. It is a close relative of the citrus that originated in China, and, as its name implies, it has a trifoliate leaf — a leaf with three leaflets.

The seeds for the rootstock are imported from Australia and harvested from trees selected especially for their general health and good root-growth characteristics. Commercial growers keep the largest seeds, air dry them and then dust them with Benomyl or a similar fungicide to prevent the formation of moulds, but this is not really practicable for the home grower.

Seedlings should be raised in spring. Sow the seeds in drills about 2.5 centimetres deep spaced in rows 10 centimetres apart. Cover the seeds with sand or pumice and then water the seedbed. The sand will allow the seedlings to emerge easily as it will not form a surface crust. Keep the soil moist at all times and free of weeds. The seedlings will emerge within four to six weeks.

Throughout the summer, protect the seedlings from the sun with shadecloth. The shadecloth will also encourage the seedlings to produce tall, straight growth. Apply light dressings of a nitrogenous fertiliser every four weeks, making sure the soil is well watered. Never apply fertiliser to dry soil as root damage may occur. Liquid fertilisers such as Dig This Organic Liquid Fertiliser are also effective.

Spray the seedlings with a combination insecticide/fungicide throughout winter. They should be planted out in nursery rows just before they begin their spring growth.

The night before you plan to lift the seedlings, water them heavily and discard any with twisted stems or malformed roots. Once lifted, shake the soil from their roots and place them under wet paper or sacking to protect their exposed roots from wind until they are planted.

Plant the seedlings out in rows 1 metre apart with 30 centimetres between each plant. Wide spacing will make it easier to weed around the seedlings and to trim branches as they grow. All side laterals must be rubbed out with a finger or pencil as soon as they appear to allow at least 20 centimetres of clear trunk before branching. The earlier the laterals are removed the smaller the scarring.

In both **grafting** and **budding**, the plant being propagated (represented by the bud) is called the **scion**, while the plant being grafted onto is referred to as the **rootstock**. A small branch with several buds on it is often called a **bud stick**.

There are two ways of joining the scion onto the rootstock of a fruit tree. A piece of a shoot from the scion variety can be joined onto the rootstock. This process is called 'grafting'. When a single bud from the scion variety is joined onto the rootstock, by a special grafting technique, the process is called 'budding'. The budding method is used for citrus because it is quicker, requires less scion material, and is usually a more successful method than grafting.

Budding is carried out as soon as the bark of the rootstock lifts easily, usually at some time from the second week of November to the end of January, depending on the local climate. Give the nursery beds a thorough watering at least two days before budding to ensure the bark will lift easily. The seedlings should have a girth of between 6 and 8 millimetres at 10 centimetres above ground level. For successful budding, scion material should also have fully formed, mature, dormant buds.

You will need a very sharp budding knife and either a roll of budding tape or some rubber budding patches. Some budding tapes, for example 'Bio-Graft', have the advantage of breaking down in sunlight after a period of about eight months. The tape will also stretch as the tree grows.

The 'T' budding method (see diagrams on page 34) is the most commonly used technique for joining the bud with the rootstock. Make a 25-millimetre vertical cut through the bark of the rootstock followed by a horizontal cut across the top of the vertical cut. Use the knife to lift out the bark at the corners of this 'T'-shaped cut to make it easier to insert the bud.

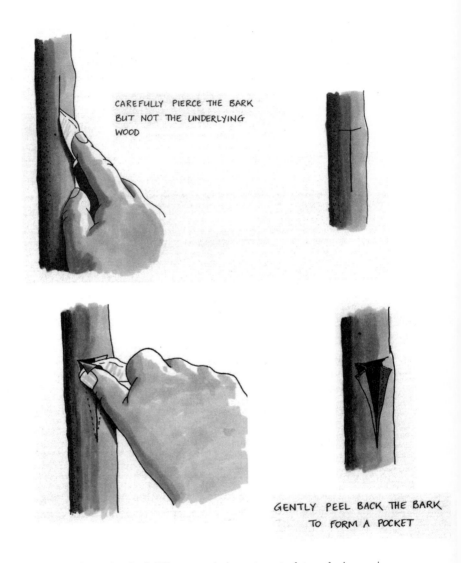

CAREFULLY PIERCE THE BARK
BUT NOT THE UNDERLYING
WOOD

GENTLY PEEL BACK THE BARK
TO FORM A POCKET

'T' budding is used to bud different varieties onto a stock tree during spring.
Make a T-shaped cut in the stock using a sharp knife as shown above. Once the
bark is peeled back the resulting 'T' cut should form a neat pocket, as
shown bottom right.

Bud sticks with plump, healthy buds are the best scions. Choose those that have had good growth during the current season, rather than those from the interior of trees that often have slender stems and small buds. Thick water sprouts that grew very vigorously are often poor scions. Cut bud sticks from the parent tree, removing the soft top and all leaves. Place the bud sticks in water to prevent them drying out.

To remove a bud from the bud stick, hold the top of the bud stick and make a sloping cut into the bud stick about 10 millimetres above the bud. Then slice off the bud by making a flat cut upwards, this time starting 10 millimetres below the bud. This slice should be perfectly straight, leaving a thin shield of wood underneath the bud. Remove this wood from the shield by tearing it or wiggling it away to expose the root of the bud. It takes practice to learn to do this properly so try experimenting with rejected buds first.

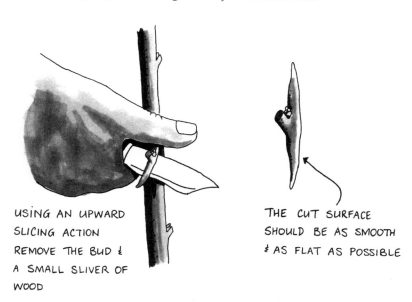

USING AN UPWARD
SLICING ACTION
REMOVE THE BUD &
A SMALL SLIVER OF
WOOD

THE CUT SURFACE
SHOULD BE AS SMOOTH
& AS FLAT AS POSSIBLE

Removing the bud.

When the bud and rootstock have both been prepared, the bud can be pushed into the 'T' cut in the rootstock, as shown below. The bud shield may have to be trimmed to fit neatly. Finally, the bud is held in place by wrapping budding tape above and below the 'T' cut.

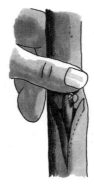

GENTLY INSERT THE BUD
SHIELD BETWEEN
THE BARK FLAPS

START WRAPPING BELOW THE INCISION:
USE A BUDDING/GRAFTING TAPE

Placing the bud into the 'T' cut. The tape will hold the bud securely in place until it has 'taken'.

If the bud is still green and healthy three weeks after the operation, then it has 'taken'. If the bud looks brown, another attempt should be made on the opposite side of the rootstock. Once a bud has 'taken', the tape can be loosened off and the plant allowed to grow unchecked until early spring, at which time all the top growth on the original rootstock should be removed, leaving the new growth from the budded section to produce the tree.

Tie the new shoot that develops from the bud to a stake to encourage it to grow upwards. When the shoot reaches 90 centimetres, remove the top shoots to encourage laterals to form. Select five or six evenly spaced lateral branches to form the framework of the tree and remove all other laterals as they appear.

Remove any flowers or rootstock shoots that appear. All the energy available should be concentrated into the new scion shoot.

Rootstock characteristics

Propagating citrus trees by budding the desired scion onto *Poncirus trifoliata* rootstock will produce a number of changes to the character of the trees that result from this union. The use of *Poncirus trifoliata* rootstock restricts the size of the tree, especially when a selection of trifoliate orange rootstock called Flying Dragon is used (see Dwarfing on page 38 for more information), but the tree will bear more fruit and it will be of a higher quality. The fruit also holds its quality longer on the tree and the trees are significantly more resistant to phytophthora rot as well as collar-rotting fungi. The trees can also resist most strains of tristeza virus when grown on *Poncirus trifoliata* rootstock and are less susceptible to damage by citrus nematodes.

Trees with this rootstock will grow better in heavier, poorly-drained soils but do not do well on alkaline soils or calcareous soils (soils with a high proportion of chalk or lime). *Poncirus trifoliata* is susceptible to scaly butt virus which is why it is not used as a rootstock for lemons; Villa Franca, Eureka and Genoa lemon varieties all carry this virus.

Poncirus trifoliata is well suited to New Zealand's colder, more marginal, subtropical climate. There are two groups of *Poncirus trifoliata*, one that features large flowers and the other with small flowers. The large-flowered group is noted for stronger, upright growth which usually results in a better, more easily managed trunk. As a result, the large-flowered group is more commonly used in New Zealand. The small-flowered group produces weaker growth, branches earlier and requires more pruning.

Because *Poncirus trifoliata* is virtually the single most important rootstock used for citrus trees in New Zealand, there is a danger that a new virus could develop which may affect all of these trees. Because of this risk, Kerikeri Research Orchard continues to carry out work on alternative rootstocks. Trials have also been undertaken with different scion varieties.

Dwarfing: A 'dwarf' tree is often defined as being from 1.5–1.6 metres tall, once it is mature, compared to a standard-sized tree, which would reach a height of 4.5–6 metres tall at maturity. A number of factors can influence tree size, but generally speaking, tree size can be effectively modified by choosing the appropriate rootstock. For example, trees on small-flowered rootstocks such as Rubidoux, are typically 15 to 20% smaller than trees on large-flowered selections, such as Pomeroy. Similarly, a common grapefruit or sweet orange tree grafted onto a dwarfing rootstock, such as Flying Dragon, is most likely to grow only to 'dwarf' size.

Dwarf or smaller trees have advantages for commercial growers, who have shown considerable interest in dwarfing rootstocks for citrus so they can plant more trees over a smaller area, while also cutting back on hedging and pruning costs. Harvesting from smaller trees is also easier.

Tree size can also be affected by a number of environmental factors, such as soil type, mild drought, soil salinity or alkalinity, marginal growing sites, or climate. Trees on trifoliata rootstock are often smaller-sized when grown in deep, sandy soils even with irrigation. On better soils, these trees grow slowly but reach standard size.

6: Citrus varieties

THERE ARE A NUMBER of different citrus varieties producing fruit that vary in size, shape, colour and flavour. Many of these varieties have been grown for centuries, whereas others are more recent developments. A major factor to take into consideration, when selecting a citrus variety, is its climatic suitability. Warm, frost-free locations usually pose few problems. However, colder districts where frost poses a threat will make the selection of early ripening varieties a necessity. If you are growing citrus primarily to provide fresh juice then the ease of peeling and a low seed content will not be a priority. People who prefer to peel and eat citrus should select varieties that provide these attributes.

Whether it has many, few, or no seeds is not a good criterion on which to base the selection of a citrus variety. Some of the best-tasting varieties produce a large number of seeds. Seediness varies from tree to tree even within the same variety. There is a range of factors involved in the amount of seed produced, with different factors influencing different varieties. Navel oranges, for example, lack seeds due to their lack of viable pollen. Sometimes cross-pollination from another variety will lead to the production of a few random seeds but usually these oranges are seedless. Parthenocarpy, which is the ability of fruit to develop without pollination, is one of the main factors that determines the production of seedless fruit. Satsuma mandarins produce some pollen, but usually remain seedless unless cross-pollinated.

General characteristics of citrus grown in New Zealand

Variety	Season	Fruit Size	Colour
GRAPEFRUIT			
Chandler	Aug–Dec	large (400 g–1 kg)	yellow
Cutlers Red	July–Aug	medium	deep red skin
Golden Special	July–Nov	medium	rich golden
Hawaiian	July–Nov	large	pale yellow
Jamaican	July–Nov	medium–large	pale yellow
Star Ruby	Aug–Dec	medium–large	pale yellow
Wheeny	Nov–March	large (440 g)	pale yellow
LEMONS			
Eureka group			
Eureka	June–Aug	medium	yellow
Genoa	Dec–Feb	medium	yellow
Villa Franca	Dec–Feb	medium–large	yellow
Lisbon group			
Lisbon	year round	medium–large	yellow
Yen Ben	June–Aug	medium	yellow
Lemon hybrid			
Meyer	year round	medium	orange-yellow
LIMES			
Bearrs	May–July	medium	green
Kaffir	year round	small	green
Key	April–Dec	small	yellow-green
Kusaie	year round	small	yellow
Mexican	June–Aug	medium	deep orange
Rangpur	June–Aug	small	reddish orange
Tahitian	May–July	medium	green
Yuzu	July–Sep	small	yellow

Pips	Flavour	Comments
few	very sweet	Pink to red flesh, better as an ornamental in NZ
many	sweet	Good bearer, highly ornamental, similar to Golden Special
few	sweet	Juicy with an excellent flavour, improved from old Morrisons Seedless
none	tart acidic	Thickish skin, requires warm climate
none	mild	Needs a warm climate
few	mildly acidic	Needs a warm climate
many	acidic	Soft skin, biennial*
		This group are virtually thornless
none	acidic	Heavy cropper, holds well on the tree
few	acidic	Thick rind, heavy cropper
none	acidic	Heavy summer crop
		This group are relatively thorny
none	acidic	Strong grower, best crop late winter
few	acidic	High yields, thin rinds, juicy
many	sweet/tart	Possibly lemon/mandarin hybrid, juicy
none	juicy	Heavy bearing, hardy
none	juicy	Leaves used in Thai & Malaysian recipes, fruit inedible
few	acidic	Frost tender, juicy, used in Key-lime Pie
many	acidic	Prolific producer, very juicy, actually an acidic mandarin
none	very sour	Easy-peel, good crops
few	acidic	Hybrid mandarin/orange and lime
none	juicy	Also known as the Green Lime or Persian Lime
few	acidic	Very thorny, highly prized in Japanese dishes

Variety	Season	Fruit size	Colour
MANDARINS			
Satsuma group			
Aoshima	May–July	medium	orange
Kawano	May–July	medium–large	orange
Miho	May–July	medium	orange
Miyagawa Wase	May–July	medium	orange
Okitsu Wase	June–Aug	medium	orange
Satsuma	June–July	small–medium	orange
Silver Hill	June–July	medium–large	orange
Others			
Burgess Scarlet	Sept–Oct	medium	highly coloured
Clementine	July–Oct	small–medium	deep orange
Corsica No. 2	June–July	largish	orange
Encore	Nov–Feb	medium–large	orange
Fortune	Sep–Nov	medium	reddish orange
Hansen Late	Sep–Nov	medium–large	orange
Kara	Oct–Dec	medium–large	reddish orange
Murcott	Oct–March	medium–large	shiny orange
Nova	May	medium–large	deep orange
Ortanique	Oct–Nov	medium	bright orange
Richards Special	Aug–Sep	large	bright orange
Sweetie	June–Aug	medium	orange
Sunset	Sep–Oct	medium	orange
Thorny	September	medium	great colour
Zest	July–Nov	medium	orange
ORANGES			
Navels			
Bests Seedless	Aug–Nov	medium	orange
Carters Navel	July–Sep	medium	orange

Pips	Flavour	Comments
none	slightly sweet	Easy-peel, small growing
none	slightly sweet	Easy-peel, medium size, heavy cropper
none	high sugar	Easy-peel, popular export variety
none	very sweet	Thin skin,easy-peel, needs warmth
none	slightly sweet	Easy-peel, vigorous grower
none	sweet	Very popular, easy-peel
none	slightly sweet	Easy-peel, very juicy, earliest mandarin to ripen
none	very sweet	Aromatic, vigorous grower
many	sweet	Heavy cropper, compact growth
few	very sweet	Heavy cropping Clementine
few	very sweet	Vigorous, juicy, thin skin
none	sweet/tangy	Susceptible to 'citrus blast'
none	sweet	Vigorous, very juicy
few	very sweet	Excellent flavour, open weeping habit
few	very sweet	Thin skin, very juicy; susceptible to Alternaria**
few	sweet/tangy	Easy-peel
few	sweet/tangy	Excellent cropper, hard to peel, susceptible to scab
few	sweet, aromatic	Short season, highly ornamental
none	very sweet	Mandarin/tangerine hybrid, juicy, high in Vit. C, easy-peel
none	sweet	Mandarin/tangerine hybrid, low acid, skin has darker stripes
many	sweet, aromatic	Also called 'Mediterranean' and 'Willow Leaf', short season, biennial cropper
none	sweet/tangy	Mandarin/tangerine hybrid, easy-peel, high in Vit. C
none	sweet	Heavy cropper, juicy, ideal for containers
none	very sweet	Very juicy, can be left on tree

Variety	Season	Fruit size	Colour
Franklin Navel	July–Sep	medium–large	orange
Fukumoto (Navel)	July–Sep	medium–large	deep orange
Johnson Navel	July–Sep	large	orange
Leng Navel	July–Sep	medium	orange
Parent Navel	July–Aug	medium–large	orange
Robertson Navel	July–Sep	large	orange
Summer Navels	July–Dec	large	pale orange
Washington Navel	July–Sep	large	deep orange
Others			
Blood Orange	July–Aug	medium	orange
Cara Cara	July–Sep	medium	orange
Cipo	July–Nov	medium	orange
Hamlin	June–Aug	small–medium	orange
Harwood Late	Nov–March	medium	orange
Jaffa	Sep–Jan	medium–large	light orange
Pineapple	Jul–Sep	medium–large	orange
Ruby Blood	Sep–Nov	medium	red-pink
Sanguinella	July–Sep	small–medium	reddish orange
Seville	July–Sep	medium	orange
Valencia	Nov–March	medium	orange
Vainiglia Pink	August	small	orange
ORNAMENTALS			
Buddha's Hand	June–Aug	large	yellow
Chinotto	Aug–Dec	small	deep orange
Kumquat	June–July	small	golden yellow/ orange
Limequat	June–Aug	small	light yellow

Pips	Flavour	Comments
none	very sweet	Excellent flavour
none	very sweet	Highly recommended, vigorous grower
none	very sweet	Good juice content
none	very sweet	Australian origin
none	very sweet	
none	very sweet	Slow-growing fruit in tight clusters
none	very sweet	Late-hanging navels from Australia, cultivars grown in NZ are 'Autumn Gold', 'Barnfield', 'Powell', & 'Summer Gold'
none	very sweet	Excellent flavour, lasts for months on tree, one of the best in NZ
few	mild sweet	Generic term for sweet oranges with red flesh, cultivars include 'Moro' & 'Tarroco'
none	very sweet	Dark pink-orange flesh
none	very sweet	Pineapple flavour, very pendant habit, can be used as ground cover
few	tart	Easy-peel, excellent juicing orange
none	very sweet	Rich flavour, very juicy, good juicing orange
few	very sweet	Similar to Valencia, strong aroma, thin peel
few	sweet	Very juicy, pebbly peel, good for juicing and eating
few	very sweet	Best tasting of all oranges
many	very sweet	A Spanish deep blood orange, lasts on tree well
many	very sour & bitter	Excellent for marmalade & preserves
none	very sweet	Rich flavour, very juicy, good juicing orange
many	mild sweet	Tastes like a melon, pinkish flesh, acid-free taste
none	none	Unusual hand-shaped fruit, grown for its aroma
many	bitter/sour	Dwarf, bitter orange, good container plant
few	acid-sweet	Juicy, good for marmalades or eaten fresh, skin and all
few	acid-sweet	Cross between a Mexican Lime and Kumquat, use as a lime substitute, juicy flesh, cool climate

Variety	Season	Fruit Size	Colour
Variegated Eureka	June–Aug	medium	green-yellow
TANGELOS/ TANGORS			
Dweet	July–Aug	medium–large	bright orange
Gold	Sep–Oct	large	deep orange
Minneola	June–Aug	medium–large	bright orange
Seminole	Aug–Oct	medium–large	red-orange
Tinura	Sep–Nov	medium	deep orange
Ugli	Sep–Oct	medium–large	bright orange

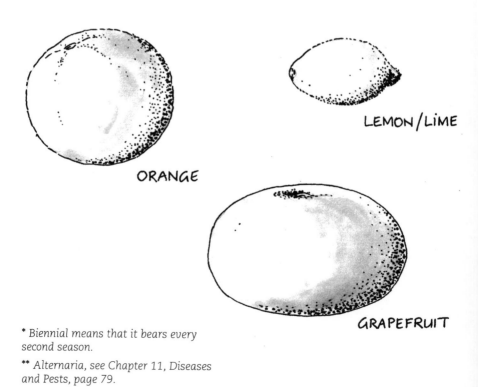

LEMON / LIME

ORANGE

GRAPEFRUIT

* Biennial means that it bears every
second season.

** Alternaria, see Chapter 11, Diseases
and Pests, page 79.

Pips	Flavour	Comments
few	tart acid	Very attractive ornamental, young fruits variegated, variation 'Villa Franca'
few	sweet	Orange/mandarin hybrid, juicy
none	very sweet	Compact grower, easy-peel, aromatic
few	very sweet	Hybrid grapefruit/tangerine, good eating fruit
many	very sharp	Rich juicy flesh, hard to peel, tender flesh
few	sweet	Good juice content, vigorous grower
few	good flavour	Rind thick, easy-peel, worth growing

TANGELOS

MANDARIN

Ease of peeling is an important consideration for many people. With the exception of lemons, limes and citrons, most citrus fruits peel easily; however, some varieties peel more easily than others. Sometimes puffiness of the rind is a sign of overripeness rather than an easy-to-peel fruit.

Another variable that is hard to determine is flavour and juiciness. Fruit technicians measure the sugar-to-acid ratio in commercial crops (known as the Brix test) along with a long list of other indicators that help them measure flavour. Soil conditions, the maturity of the tree, climate and seasonal weather fluctuations can all influence the flavour of citrus fruit. Different varieties will also have different flavours, but remember that other factors also play a part in how the fruit will taste.

By carefully selecting different varieties of citrus, the harvest of fresh fruit can be extended throughout the year. Annual differences in weather patterns and local microclimates will influence when fruit will ripen. The best way to determine when the fruit is ripe is to pick one and taste it. If it is bitter the fruit needs to be left longer on the tree. If it is dry then the fruit has been left on the tree too long.

Oranges

Sweet oranges

The sweet orange is the most important member of the citrus family. Historically, oranges are believed to have originated in north-eastern India, Burma and China. Introduced to Europe by 300 AD, they had reached America by 1518. Today, nearly all oranges grown in New Zealand are varieties of the sweet orange. By planting a selection of varieties, oranges can be harvested from August until after Christmas.

Navel oranges

Navel oranges are renowned for their crisp, rich flavour and ease of peeling. They take their name from the presence of a small secondary fruit that is embedded in the apex of each fruit, creating the appearance of a navel. Navel oranges grow well and produce their best flavour only under a narrow range of cool climatic conditions, but the fruit usually ripens early enough to allow harvesting before the worst frosts occur. The trees are slower growing than other varieties, such as 'Valencia', and the yields are lower but the trees can be closely planted in small groups to compensate for this.

Navel oranges are seedless and produce their fruit between August and September. The flesh is deep orange, of good, firm texture and moderate juice content. These fruit characteristics make navel an ideal fruit for the fresh fruit market and in New Zealand, when budded onto *Poncirus trifoliata*, they can be of the highest eating quality.

Navels grown in New Zealand have traditionally been harvested in August and September, but in a good situation they can be picked in July, and on *Poncirus trifoliata* rootstock some strains of navel orange will stay on the tree until the end of October. Grown on sweet orange rootstock, navel oranges tend to dry out quickly by the end of September.

'**Washington Navel**' is one of the most important varieties and is the standard eating orange. It was introduced to New Zealand last century. It is a medium-sized tree, with a rounded top and a slightly drooping habit with dense, dark green foliage. Fruit position is intermediate; that is, the fruit are well spread through the whole canopy of the tree, rather than clustered on the outside or the top. The fruit is large and deep orange in colour, but it is very tight-skinned and therefore not easy to peel. The fruit holds on the tree for two to three months and harvests from August to October. 'Washington Navel' is the world's main commercial strain and has been the source of a number of other commercial strains.

'**Parent Navel**' originated from one of the three original 'Washington Navel' trees in California. It is a moderately vigorous tree for a navel, producing good-sized fruit of good quality that can hang well on the tree until late in the season. It is as good as any strain tested in New Zealand and is likely to be the predominant strain in New Zealand orchards in the future.

'**Carters Navel**' is one of the most popular sweet oranges for home gardeners. Both the tree and the fruit are virtually indistinguishable from 'Washington Navel'; however, the rind of the 'Carters Navel' is smoother and thinner and the flesh is softer, juicier and sweeter, but the skin is tight and difficult to peel. It also matures slightly earlier.

'**Johnson Navel**' is the main variety planted in Northland, and originated from the trees imported from Australia. It has a large fruit with good juice content.

'**Leng Navel**' is also of Australian origin and produces a smaller, earlier-maturing fruit than 'Washington Navel'.

'**Robertson Navel**' is a slow-growing tree even for a navel, but produces large fruit, often in tight clusters.

Other strains of navel orange grown in New Zealand include '**Bests Seedless**', '**Franklin Navel**' and '**Fukumoto**'. '**Bests Seedless**'originated in Avondale, Auckland. More vigorous than other navels, it is a good cropper on trifoliata rootstock. Performance on other stocks is more variable. The fruit is medium in size and matures from August to November. However, fruit size can be a problem and there is a marked tendency to biennial bearing.

Common oranges

Common oranges include **Valencia**, **Hamlin**, **Pineapple** and **Jaffa** varieties. Valencia oranges are the most important variety of common orange and are well established in all the other countries that grow citrus fruits. They are vigorous trees that produce high yields of moderate-sized fruit. New Zealand is a marginal growing area for Valencia oranges because of their heat requirement. The fruit often fail to reach full size when grown under New Zealand conditions. For this reason, Northland is likely to be the best area for them. The fruit is not always well coloured as they are a late-season orange and tend to suffer from re-greening in hot conditions at harvest time. However, grown on trifoliata rootstock, Valencia oranges can be left on the tree late into the season without drying out as they have high juice content.

'**Harwood Late**' is a strain of Valencia orange that originated in Tauranga as a seedling from a good Valencia tree. The fruit is indistinguishable from the parent tree and it has become the only important strain of Valencia in New Zealand. Grown on trifoliata rootstock, the tree is large, vigorous and productive.

There are no other oranges of commercial significance in New Zealand, but other varieties can occasionally be found.

Pigmented or blood oranges

Blood oranges, known also as pigmented oranges, take their name from the pink or red coloration of the flesh, on the rind and in the juice. They have a distinctive flavour and are regarded by many as the best tasting of all oranges. The only variety grown in New Zealand is '**Ruby Blood**'. The reddish, slightly oblong fruit is produced on a medium-sized, compact, moderately vigorous tree which matures mid-season. '**Sanguinella**', a Sicilian variety, is being tested in Kerikeri.

Sour oranges

These are also known as bitter or '**Seville**' oranges, and are a native of the north-eastern regions of India, parts of Burma and China. There are a number of similarities between sour oranges and sweet oranges but there are also important differences which clearly justify their separation into different species. The sour orange leaf is darker in colour and more tapered than the sweet orange, and the petiole (the stalk that joins a leaf to a stem) is longer and more broadly winged. It is a more upright and thorny tree, and much more resistant to frosts, excess moisture and neglect.

Apart from the obvious difference in flavour, the fruit is usually flatter and more deeply coloured and the rind is thicker.

Sour oranges are too bitter for most people's tastes, but can be used to make a distinctive and refreshing drink. The main use for sour oranges is to produce marmalade and the liqueurs Curaçao and Cointreau.

Mandarins

Mandarin trees are the most cold resistant of the citrus trees. Their fruit, however, is thin-skinned and small in size which makes them more susceptible to cold damage than the larger oranges and grapefruit. The mandarin probably originated in north-east India and historically has been of major importance in the Orient.

Mandarins grow better when the weather is hot and atmospheric humidity is high. Fruit size is larger under these conditions and the fruit also becomes juicier and milder due to a lower acid content. Mandarins usually only last for a few weeks on the tree, after which time the rind 'puffs' and both the acidity and the juice content diminish rapidly.

Satsuma

This mandarin is the famous and highly important *unshu mikan* of Japan. The name Satsuma, by which it became known, is credited to an American woman who sent trees home from Japan in 1878, Satsuma being the name of the province where they are believed to have come from. These mandarins probably originated from a seedling introduced from China prior to the sixteenth century.

In the early days of modern Japanese horticulture there were five kinds of *unshu mikan*. Between 1908 and 1911, over a million Satsuma trees were exported to the United States and by 1932 30 varieties were named. This list now extends to over a hundred varieties.

In New Zealand, a research programme began at the then Kerikeri Horticultural Research Station in 1983, with the aim of comparing 47 citrus cultivars. The programme was extended in 1986 with 10 Satsuma mandarin varieties included. Satsuma mandarins are currently the second most important mandarin in New Zealand, after '**Clementine**'. They are slow-growing, somewhat spreading trees and small compared to most other citrus. The leaves are usually dark green and the tree habit is very open. They are among the most cold-tolerant citrus grown in New Zealand.

The fruit often matures before it reaches full coloration and the skin is very easy to peel. They are seedless and eating quality is fair. They ripen early, starting before the end of May, with the season going through to late July or early August.

Two of the early strains of Satsuma seen in New Zealand were '**Silverhill**' and '**Wase**', but as growers bought trees and sold fruit all simply named as Satsumas, the individual strains tended to become lost. In more recent years, some of the better Japanese strains have been introduced and these may prove to be superior for New Zealand conditions.

'**Miyagawa Wase**' is the predominant strain in Japan. In New Zealand it is the best known and most extensively grown of the 'Wase' varieties. Commercial plantings in Gisborne have been producing for several years. The fruit is large for a Satsuma, seedless, with plenty of juice and a rich flavour. The fruit also stores well. The harvest period is late April to May.

Common mandarins

'**Clementine**' mandarin is the main variety grown in New Zealand at present. Well suited to cooler growing conditions, the fruit has a low heat requirement to reach maturity. The tree is vigorous and dense with a rounded top and long, very pointed leaves. The fruit is small, with a deep orange coloration of the skin and flesh. The skin is of medium thickness and peels easily. The fruit is juicy but can become seedy if cross-pollination takes place. The 'Clementine' season runs from June to August, but on sweet orange rootstock the fruit will dry out extremely rapidly as the season advances. The trees have a biennial bearing habit.

Another important commercially grown mandarin, '**Burgess Scarlet**', is a vigorous, dense, upright tree. The fruit is small but of a high quality and easy to peel. This is an Australian variety that originated in Queensland in 1908 from a seedling grown by E.A. Burgess. The tree has a definite biennial bearing tendency and the fruit ripens late — between August and September.

Another variety that is late-ripening and has a rich flavour is '**Kara**', a Satsuma hybrid created at the University of California Citrus Research Centre in 1915. The fruit matures between October and December, holding well on the tree. One of the main drawbacks is that 'Kara' trees are extremely susceptible to verrucosis disease (see page 82). The fruit is medium to large, and although the rind is rather thick it peels fairly easily. The flesh colour is deep orange and the fruit tender and juicy with a rich, distinctive flavour. The trees are moderately vigorous, spreading and round-topped, similar to a Satsuma.

'**Encore**' is the result of another cross made between different varieties of mandarin by researchers in California. It was introduced into commercial production in 1965. 'Encore' mandarins mature late, between October and January, and hold well on the tree. The trees are thornless, moderately vigorous and bushy. However, they are somewhat 'biennial' in cropping. The fruit is medium-sized with a thin, brittle rind that fades to yellow as it matures. The flesh is deep orange, firm in texture, tender and juicy, but it does have numerous seeds.

'**Fortune**', another variety that bears regular crops of medium-sized fruit, produces fruit from early September to late November. The skin is reddish orange, thin and tends to be adherent. The eating quality is good, but the trees appear to be susceptible to 'citrus blast'.

'**Murcott**' is a moderately vigorous variety with a very upright growth. It does, however, have a high nutritional requirement. The trees require more potassium when the crop is maturing between November and February during its 'on' or heavier fruiting year (in comparison to its 'biennial bearing' or 'off year', see Chapter 4, Nutrition, page 26). This can be applied as an extra dressing of sulphate of potash, at 28 grams per metre, well watered into the soil. The fruit is small to medium in size, with a shiny orange colour, but it does have a tendency to re-green around the stem end. The skin is thin and brittle and tends to adhere to the fruit. The flesh has a rich flavour but unfortunately is somewhat grainy or sandy in the texture.

'**Nova**' is a cross between the 'Clementine' mandarin and 'Orlando' tangelo. It is a moderately vigorous tree with good crops of medium to large fruit. This is a very early ripening variety with the season running from May until July. The fruit is not easy to peel, especially straight off the tree. When peeled fresh, the peel exudes an acetylene-like smell which is objectionable to some people, but the smell is not noticeable as the skin withers. The fruit breaks into sections readily and has a dark orange colour. The flavour is pleasant but the seeds are numerous in mixed plantings (when many varieties of cultivars are planted within close proximity to each other). For example, if 'Meyer' lemons are planted near so-called seedless grapefruit, bees will cross-pollinate resulting in more seeds than usual in the grapefruit.

An Australian variety that has been grown in New Zealand for a number of years, '**Richards Special**' is notable for its resistance to a number of rind-blemishing fungal diseases. It is a vigorous, upright tree which bears large, flattish fruit. The fruit combines a high sugar level with a high acid content which imparts a rich flavour. The fruit is easy to peel. The harvest period lasts about four weeks and while the bearing habit appears to be regular, the yields are only moderate. In cooler areas, both growth and fruit size are diminished. In Kerikeri, fruit begins colouring in July with the fruit maturing in late September to early October. In Poverty Bay, maturity is later; from early October to early November. Some of these maturity differences may be related more to differences in seasons in which fruit were tested rather than to actual site or climate responses.

At several locations around the country, 'Richards Special' appears to have a harvest period of approximately four weeks, during which the internal quality is highly acceptable. Fruit tends to puffiness after this time. The combination of relatively high sugar and acid levels imparts a rich flavour. This is added to by distinct aromatic elements, resulting in a very pleasant taste.

Tangelos/Tangors

Tangelos are a group of citrus that originated as the result of a controlled breeding programme in Florida in 1897. Grapefruit and mandarins were crossed to produce a number of varieties of tangelo. The only variety of importance in New Zealand is '**Seminole**', which originated in Florida and is a cross between 'Duncan' grapefruit and 'Dancy' mandarin, with characteristics that are midway between those of its parents.

The fruit is large with high juice content. The skin is hard to peel and the fruit contains a number of seeds so they are more suitable for juicing than eating fresh. Because the fruit of 'Seminole' tends to split at the blossom end, it is susceptible to fungal infection. The fruit ripens between September and November and has an excellent flavour. The trees are capable of withstanding heavy frosts and should be grown in well-drained soils.

'**Tinura**' tangelo is more vigorous than 'Seminole' but the fruit tends to dry out quickly at the end of the season. The juice content is high and the fruit medium-sized with the flavour tending more to the grapefruit side. 'Tinura' originated in Rarotonga.

'**Minneola**' tangelo is another variety that originated in Florida. The tree is vigorous and will produce good crops of medium-sized fruit. The flavour of the fruit is excellent and its seed content is low. 'Minneola' tangelos mature from September to October.

'**Ugli**' tangor also ripens from September to October and the large fruit is tender and juicy, with a good flavour and few seeds. The rind is thick and coarse which allows Ugli tangor to be peeled easily. The rather unusual name for this natural tangor is reputed to have been given to this unattractive but delicious fruit in the Canadian market which first received it. 'Ugli' probably developed from a chance seedling of unknown parentage in Jamaica, and while the fruit is unattractive, its shipping and eating quality have given it a high reputation in Canadian and English markets.

Grapefruit and pummelos

Grapefruit and pummelos exhibit so many resemblances that their close botanical relationship is obvious. The main variation is in fruit size, form, and rind thickness. Pummelos are the largest of all citrus fruits and range in flavour from highly acidic to insipidly sweet, whereas grapefruit exhibit a uniform pleasant and distinctive flavour.

The main pummelo grown in New Zealand is known as the Tahitian grapefruit. This is a highly juicy pummelo with a thin rind, thought to have developed in Tahiti from seed that came from Borneo.

'**Chandler**' pummelo is another variety with large, pleasant-tasting fruit. 'Chandler' is under trial at the HortResearch orchard in Kerikeri.

The true grapefruit, which are varieties of Citrus paradisi, have high heat requirements and are unable to be grown in New Zealand. There are, however, two grapefruit-like types that can be grown, the **New Zealand grapefruit** and '**Wheeny**' grapefruit. The New Zealand grapefruit was introduced about 1855 and eventually became the most important of the grapefruit types grown in commercial orchards around the country.

The grapefruit tree is vigorous and, under favourable conditions, is one of the largest citrus trees, requiring more space than any other.

With a low heat requirement and good tolerance to colder temperatures, the New Zealand grapefruit produces good yields over a long season stretching from May to January. The skin is thick but the fruit has high juice content. If grown near 'Meyer' lemons, 'Seminole' tangelos, 'Clementine' mandarins or 'Wheeny' grapefruit, cross-pollination is likely to take place which will result in seedy fruit. The New Zealand grapefruit has a definite biennial bearing tendency.

The best quality fruit are obtained when the trees are grown on trifoliata rootstock, while the best selection seems to be '**Golden Special**', which comes from a commercial orchard in Tauranga, an improved variety from the old favourite 'Morrisons Seedless'. The fruit is medium in size with few to no seeds. The flesh is tender and very juicy with a good flavour. It holds well on the tree and also stores well.

'**Wheeny**' grapefruit mature between November and February. The fruit is large, moderately seedy and very juicy. 'Wheeny' has a pronounced biennial bearing habit, producing negligible crops every second year. Of minor commercial importance in New Zealand, 'Wheeny' was introduced from Australia in 1935. The variety originated as a chance seedling at Wheeny Creek in New South Wales and is probably a pummelo hybrid.

Lemons

Lemons share a number of distinctive characteristics with limes and citrons. They are all more or less ever-flowering and ever-bearing, while the fruit is highly acidic with an oval to elliptical shape. With the exception of 'Meyer' lemon, all members of this natural group are highly sensitive to cold weather. Citrons are used overseas in the candied-peel industry and are almost unknown in New Zealand.

The lemon originated in the north-eastern regions of India and by the end of the second century had reached Italy, having been taken to the Mediterranean and across North Africa to Spain by the Arabs. The Crusaders took it to southern Europe in time for inclusion as one of the fruits taken by Columbus on his second voyage to the New World.

'**Eureka**' is a lemon variety that originated in California, grown from fruit of Italian origin in 1858. The trees are medium in size and vigour, virtually thornless and highly productive. Because the fruit is produced at the end of long branches it is sensitive to cold. The fruit is medium to small and the seed content variable, but usually there are no seeds. The flesh is tender, juicy and the flavour highly acidic. The crop is distributed throughout the year but occurs mainly from late winter to early summer.

'**Genoa**' is a Californian variant of 'Eureka' that crops throughout the year but produces most of its fruit in winter. Smaller and denser than 'Eureka', 'Genoa' trees are almost thornless and the fruit has few or no seeds.

'**Lisbon**' lemon trees are large, upright, spreading and vigorous. They are thorny with dense foliage and resistant to adverse weather conditions. The fruit is very juicy and acidic with few seeds and is produced mainly in winter and early spring. The 'Lisbon' lemon is of Portuguese origin and gave way to the 'Villa Franca' variety as the predominant commercial variety in New Zealand in the 1950s.

'**Yen Ben**' is a 'Lisbon' strain originally selected in Queensland. It has attracted commercial interest in recent years with sizeable plantings being established in Northland and the Bay of Plenty. The smallish fruit are thin-skinned.

'**Villa Franca**' lemons are virtually indistinguishable from 'Eureka', although their seasonal distribution is more like that of 'Lisbon'. Believed to be of Sicilian origin, the trees are less thorny and more open than 'Lisbon'.

'**Meyer**' lemons were introduced to the USA from China in 1908. The trees are small to medium in size, moderately vigorous, hardy and productive, with the additional advantage of being virtually thornless. The fruit is juicy, acidic and moderately seedy. The crop is distributed throughout the year but occurs mainly in winter.

Limes

As with the citron and the lemon, limes are believed to have originated in north-eastern India, Burma and northern Malaysia, and they followed the same routes westward. There are two main groups of limes, the acid or sour limes and the sweet limes.

The **West Indian lime** is the main acidic variety and is notable for having the highest percentage of acid in the juice of all citrus fruits. It is lower in ascorbic acid and vitamins, however, than the lemon. The fruit is very small and moderately seedy, with a tightly adherent skin. At maturity, the skin colour is greenish-yellow, following which it drops from the tree. The flesh colour is the famous 'lime green' and the flesh is tender, juicy and fine-grained.

The trees are small (2.5–4.5 metres) and like the lemon have irregular branches, but it is a little more bushy and spreading. It is also densely armed with small, slender spines. The trees are highly productive, producing most of their fruit in winter.

There are both small-fruited and large-fruited acid limes, and the **Tahiti lime** falls into the latter category. Tahiti lime crops mainly in winter and the yield can be poor as a considerably warmer climate is needed for the tree to perform efficiently. Tahiti lime was introduced into California from Tahiti between 1850 and 1890. It was introduced to Australia from Brazil as the Persian lime, even earlier, around 1824.

The exact origins of the Tahiti lime are unknown. It is thought to be a hybrid and that one parent is either a citron or a lemon. These limes are acidic, very juicy and lacking in seeds.

The **Rangpur lime** is another large-fruited acid lime, it is hardy to cold, vigorous, productive and has few thorns. The tree is medium-sized, spreading and drooping with dull-green foliage. The fruit holds well on the tree and the flesh is orange-coloured, tender, juicy and strongly acidic. The name Rangpur lime is confusing because it is actually a sour mandarin rather than a true lime.

Lemonades are a type of citrus that require the same care as limes. They are not very popular in New Zealand as they are particularly frost tender and need protection.

Kumquats (Fortunella)

The kumquat is not a true citrus — although the fruit looks like an orange, the rind is edible. The fruit, which is slow to colour and ripen, can be round or oval and has a delicious perfume. They are grown extensively in Australia and have been grown in Japan for centuries. Kumquats are small trees, hardy and cold resistant. They flower in late spring and summer and so miss the frosts.

Fortunella japonica, or the **Marumi kumquat**, which is still sometimes known as *Citrus japonica*, has round, golden-yellow fruit, about 2.5 centimetres in diameter, with a thin, waxy skin. It tastes acid-sweet with a juicy flesh. *Fortunella margarita*, or the **Nagami kumquat**, has fruit about 4 centimetres long and 2 centimetres in diameter and is more orange-coloured when ripe than *F. japonica*.

Kumquats make ideal pot plants and can be situated either outside as patio plants or indoors. Because all parts of the plant are sweetly scented, they make pleasant indoor companions when placed in well-lit rooms.

7: Selecting and planting

SELECT TREES WITH vigorous top growth when buying from a nursery. Trees that have been grown for one or two years after being budded are usually the best choice as they are well established and the union of the bud and scion has matured and should not break off. Don't fall into the trap of buying extra-large four- or five-year-old trees. These will not transplant as well as younger, smaller trees which are nearly always able to adjust more easily to the shock of transplanting. Younger trees will usually outgrow large older trees and come into full bearing sooner.

Avoid buying trees with any signs of hard, stunted growth, pests or diseases. If any fruit has formed on the tree, remove it before transplanting to give the tree a better chance of becoming established and building a sturdy framework capable of carrying large crops as soon as possible.

Trees budded onto trifoliata rootstock are often smaller and slower growing than other specimens in a nursery, but, provided these trees are not 'runts', or really small and unhealthy looking, this is no disadvantage as they will quickly become established and bear earlier and more heavily than those on other stocks. All citrus, except lemons, are now budded onto trifoliata rootstock.

Always select a warm, sheltered and sunny site for citrus trees, as trees exposed to prevailing winds will never succeed. Buildings and fences are better forms of shelter than hedges and shelter trees. The invasive root systems of other plants compete with the citrus for food and moisture. Also, many shelter trees harbour a number of pests and diseases.

Citrus trees must be given ample room for development. Grapefruit and 'Lisbon' lemons develop into large trees and need plenty of space; conversely, mandarins and navel orange trees grown on trifoliata rootstock can be closely planted. Trees on sweet orange or rough lemon roots require more room. Distances usually range from 3.5–5.5 metres depending on the variety, rootstock and soil conditions.

Citrus trees can be planted from April through to September. In warm regions you can plant in autumn, or even mid-winter, but in colder areas planting should be delayed until after the danger of frosts is past. If your trees have been grown in black polythene bags you can safely plant them right through the summer months, as long as care is taken to supply extra water.

If the trees have been 'balled' or wrapped in sacking and new roots are showing through, it is better to position the tree in the planting hole with the sacking in place (see diagram opposite). Cut away as much of the sacking as possible, leaving the remainder to rot in the ground. If your tree is in a polybag, cut away the bottom of the bag with a knife and carefully place the tree in the planting hole, then cut away the rest of the polybag.

It is important to make sure that citrus trees are not set too deep otherwise they may be killed by collar rot, a fungus disease which frequently develops where soil comes into contact with the bark, especially near the bud union. Plant the trees so the uppermost roots are branching out at, or even slightly above, soil level. Settle the soil firmly around the roots with your foot, or water in place if the soil is dry. A liquid fertiliser such as Dig This Organic Liquid Fertiliser can be used to give the newly planted trees a boost.

If a prevailing wind is a problem you can protect the young trees by placing three or four stakes out from the tree and tying some windbreak material around these stakes. See Chapter 8, page 67, for information on pruning young trees.

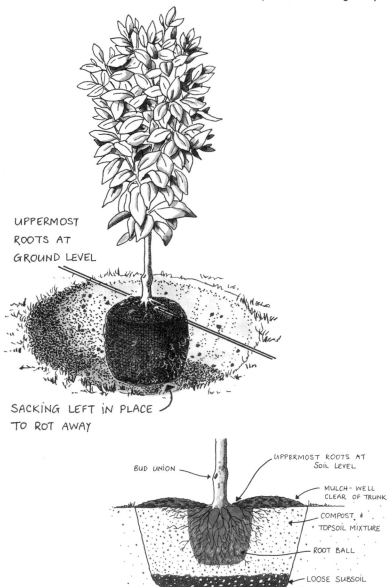

UPPERMOST
ROOTS AT
GROUND LEVEL

SACKING LEFT IN PLACE
TO ROT AWAY

BUD UNION

UPPERMOST ROOTS AT
SOIL LEVEL

MULCH – WELL
CLEAR OF TRUNK

COMPOST &
TOPSOIL MIXTURE

ROOT BALL

LOOSE SUBSOIL

*Plant your new citrus tree carefully, making sure you leave sufficient space for it
to grow to its full size. If your tree has been grown in a black polybag, cut the bag
from top to bottom with a knife and remove it completely before planting the tree.*

Recommended planting methods

Preparation

Prepare a hole at least 1.5 to 2 times the size of the container. Loosen the soil at the bottom of the hole and backfill the bottom of the hole with good soil and compost. If the soil conditions are heavy (clay) it may be best to mound up the site slightly to stop water ponding around the trunk.

Planting

Remove the tree from its container. A bag is easily removed by either cutting away the bottom of the bag and removing the rest of it once the tree is in place (as described above), or just by cutting down two sides with a sharp knife, taking care not to lose too much of the soil as you discard the bag. Remove boxes by splitting the ends — the four sides and bottom are then easily separated from the root ball.

Whatever method you use, try not to disturb the root ball. Roots should only be lightly teased out, or not at all. Backfill hole with good soil and compost flush with the top of the root ball only. Do not build soil up higher than planted around stem. The finished height for the tree should be flush with the existing ground level, or if the surrounding soil level is being built up, flush with the new ground level height.

Remove the label from the tree as this will cause strangulation as the tree girth increases and the label tie cuts into the bark.

Staking

This is essential. Drive two stakes into the ground close to the tree but avoiding the root ball. Secure citrus firmly using tree ties. Stakes and ties should be left holding the tree for at least one to two years after planting until the roots are established. Ties should be checked regularly during the growing season to ensure they are not cutting into the tree.

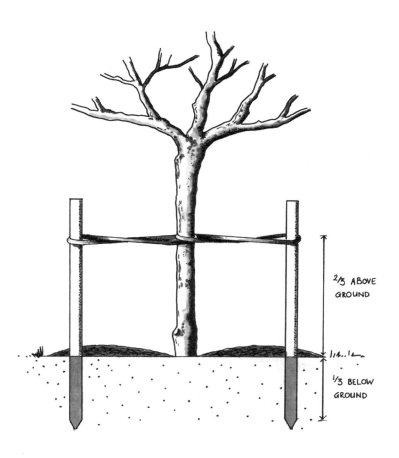

2/3 ABOVE
GROUND

1/3 BELOW
GROUND

DOUBLE TIE
VIEWED FROM ABOVE

Staking and tying a new tree.

Watering

Water immediately after planting and at least three times a week with at least three buckets of water each time. You will need to water more often during dry spells and days of high wind.

Mulching

On a hot day 1 square metre of bare soil can lose up to 2 litres of water. A mulch of compost around the trunk — but not touching the trunk — will help to retain moisture. Cover the root zone area with 75 millimetres to 100 millimetres of the mulch. This will assist with water retention and also reduce weed growth.

# 8: Pruning

GENERALLY, IT IS BETTER to avoid pruning healthy foliage or interfering with the natural growth habit of most citrus trees. However, lemons and dense, bushy mandarins benefit from careful pruning.

Citrus trees grow naturally into well-shaped trees, but in formative years watch for any shoots appearing below the bud union, which is clearly visible as a slight knuckle or swelling on the trunk, near ground level or just above. Any growths appearing below this union (see diagram on page 68) must be removed as soon as they are noticed as dormant buds which have sprouted on the rootstock will give rise to unwanted rootstock growth. Growth from the trifoliata rootstock is recognisable by its smaller leaves and spiny thorns. Remove this growth by pulling in a down and outwards motion. This will also remove any dormant buds which are usually left if the branch is cut off. Use a sharp knife to clean up the tear and then treat the wound with a pruning paste containing a fungicide.

With young trees it is important to prune them back to three or four main branches immediately after planting, removing all growths that tend to grow into the centre of the tree and all spindly or weak growth.

The main branches can be pruned back by one-third to an outward-facing bud. I remove all dormant buds on the inside of branches with a sharp knife. Remove all flower buds and fruit on young trees and let them develop a strong canopy before allowing fruit to set. This usually takes about a year.

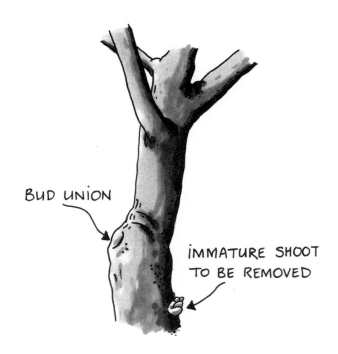

BUD UNION

IMMATURE SHOOT
TO BE REMOVED

A bud union with an immature shoot appearing below it. Remove the immature shoot by rubbing it off.

After this initial pruning, dozens of dormant buds will 'shoot' and these should be removed by literally rubbing off as soon as they are noticed. Otherwise, all your good intentions of forming a good framework will come to nothing.

As citrus trees grow to maturity, some branches may grow more vigorously than others giving the trees an unshapely appearance. Rather than waiting until this occurs, the overly vigorous shoots can be removed at an early stage. Commercial growers also prune mature trees to simplify harvesting, spraying and other operations. They also often prune to ensure that light can reach the fruiting areas. This must be done with care because a citrus tree produces food in its leaves. The amount of foliage on the tree will have a direct influence on the amount of fruit the tree will be able to produce.

CUT JUST ABOVE AN
OUTWARD-FACING BUD

The correct way to prune a lateral.

When pruning becomes necessary, the correct procedure is to thin out laterals along a branch, or cut branches back to a lateral so that good strong terminal growth always remains. A lateral is a small shoot or branch growing from the side of a main stem. (See diagram above.) Avoid leaving stubs when removing large branches.

Always make the cuts close to the main branch and seal pruning wounds with an insecticidal paint. Lemon-tree borer can gain entry when the bark is broken and lay eggs in pruning wounds.

You can also prune old or weak trees to give them a new lease of life. The degree of pruning will depend on the condition of the tree. If a tree has been badly infested by borer it may be past the stage where pruning will be effective, but if it has simply grown too large, pruning will certainly help. Study the basic framework of the tree. Remove all dead and diseased wood, then cut back all of the branches except for the main scaffolds, leaving almost no foliage.

KEEP LOW-GROWING
BRANCHES CLEAR OF THE GROUND

– BUT – OVERPRUNING
WILL RESULT IN LOSS OF QUALITY FRUIT
FROM LOWER BRANCHES

Correct pruning will result in a well-shaped, healthier tree and better fruiting.

All cuts should be sealed with pruning paste, which will need to be reapplied until the scar tissue grows over the wound. A coat of white latex paint will protect exposed branches and bark on the trunk during summer in hot areas until new foliage grows.
Extra feeding will help the tree to produce healthy new growth. Prune old trees in spring when the danger of frost has passed. Younger trees can be pruned most conveniently when the fruit is being harvested.

If severe pruning is necessary to renovate neglected, debilitated trees this should be done in mid-spring when the danger of frost has passed, growth is at its peak and wounds will heal more quickly.

Keep low-growing branches clear of the ground, but pruning to raise the 'skirt' of the tree should not be excessive, as a large proportion of the best fruit is produced on this part of the tree.

Once orange and grapefruit trees begin bearing, they normally require very little attention, other than pruning out any water shoots and the removal of dead and diseased wood. Water shoots develop from dormant lateral buds on the trunk and have very long and variable lengths between the nodes of the stem.

Lemon trees, including 'Meyer', and dense, bushy mandarin varieties will benefit from a moderate thinning of the fruit-bearing wood each year. This reduces the number of fruit the tree produces but will improve their quality and size. 'Clementine' mandarins benefit from this pruning; however, Satsuma varieties and 'Kara' require little pruning and should be treated in the same way as oranges. Some lemon varieties, especially 'Eureka', should have their branches shortened as they tend to grow too long and spindly. These branches are weak and if left to grow long are prone to breaking under the weight of fruit.

Frost damage

The most important thing to do when trees are damaged by frost is to wait until the extent of the damage can be seen. As new growth appears, the damaged areas will become clearly visible. This may take three to four months.

Symptoms of minor frost damage are yellow, droopy and wilted leaves that turn brown and fall off. Frozen leaves may not change colour but may just shrivel and die before they drop. In some cases, the leaves dry and shrivel on the twig and remain on the tree for several weeks.

The naked, dead twigs should be removed along with any small branches that have been damaged. Always wait until the danger of frosts has passed and new growth has begun before pruning. Frost-damaged growth will give some protection to new growth trying to sprout below.

 **9: Harvesting**

DON'T EXPECT TO HARVEST much fruit during the first two or three years after planting. If there are fruit on the tree when you buy it, remove them at planting time. Also remove any fruit that does appear during the first year or two. This allows the tree to use all the food it manufactures to help produce a strong framework which will support heavy crops later. In the first few years the aim should be to grow a big, vigorous tree.

A young, vigorously growing tree will tend to produce flowers within 12 months of the initial basic pruning. Remove two-thirds of the flowers; all other flowers should be allowed to form fruit. This fruit should then be removed while still small because the young branches tend to bend downwards if they bear heavily. This can be done in January or February.

Citrus fruit does not continue to ripen once it has been removed from the tree. In New Zealand, the fruit may be fully coloured but still far from mature, so care should be taken to avoid picking it too early. Check the table in Chapter 6, pages 40–47, for the correct harvesting times of different citrus varieties.

Generally, the earliest sweet oranges and New Zealand grapefruit required for dessert eating are not properly ripened until August. Other varieties ripen even later. Only certain mandarins, such as 'Clementine' and Satsuma varieties, are ready earlier.

Most citrus continue to improve in quality on the tree for a considerable time after they first mature for eating. If left on the tree too long, however, they tend to lose their juice and dry out.

The definition of maturity in lemons is rather vague and is

generally associated with the availability of extractable juice, regardless of the size and colour of the fruit.

Lemons can be harvested throughout the year and are usually picked when they are 7–8 centimetres in diameter. 'Lisbon' and 'Eureka' lemons will become oversized and coarse if left on the tree too long. This can also stop them from curing, a natural process whereby the skin thickness reduces and the juice content increases after harvesting.

If you want to store fruit for later use, the secret of success is to harvest and handle the fruit with the utmost care. Even the slightest injury to the skin will provide an entrance for decay organisms. Fruit can keep for many weeks if it is in good condition.

Citrus fruit should never be pulled from the tree. The peel may be torn next to the stem button and become infected with moulds. Also, if the tree is damaged by the tearing and breaking of twigs, it will become open to disease and borer. Remove the fruit with sharp secateurs by making two cuts. The first cut takes the stalk back to a healthy new shoot and helps to prune the tree, and the second cut removes the stalk close to the stem button. By 'pruning' the tree in this way the tree is kept compact and fruitful, and reasonably free of dead and worn-out fruiting wood.

Delay picking for several days after heavy rain as the fruit should be thoroughly dry at harvesting. Store any fallen fruit separately and use it first. Inspect stored fruit from time to time and always remove any showing signs of decay. Avoid dropping the fruit to prevent bruising, and store it in a container which is dry, free from soil, nails or splinters — indeed anything likely to injure the fruit.

10: Growing citrus in containers

IN SEVERE CLIMATES, where winter cold restricts the growth of citrus trees, gardeners can still successfully grow them in containers in sheltered places outside in the warmer months, on patios for example, and then move the trees inside during the coldest months, enabling them to survive the harsh winter conditions. Many people keep their citrus indoors year round, treating them as they would any other houseplant. All citrus budded onto trifoliata rootstock are suitable for indoor culture.

There are not many places in New Zealand, however, where citrus cannot be grown outdoors except certain districts where winter temperatures drop continually below 5°C. But there is no reason why, in these areas, they cannot be grown in containers and moved into a more sheltered spot for the winter, as described above.

Many gardeners in the South Island and inland areas of the North Island have small greenhouses where container-grown citrus can be kept safe during the winter. Ninety per cent of all citrus are now grown on trifoliate rootstock which gives them a reasonable degree of cold-hardiness and because it suits container growing due to its smaller growing habit.

As with any plant grown in a container, one of the most important things to consider is the quality of the growing media you use. Some plants only grow well in certain soil mixtures, but citrus will grow in a wide range of soil types. However, don't use soils dug from your garden as these contain weed seeds and possibly disease organisms which may be harmful to your plant. Ideally, use one of the commercially available potting mixes which are clean, lightweight and easy to handle.

It is critical that you use a suitable mix. Most good garden centres will have tree and shrub mixtures developed especially for growing citrus in containers, and the garden centre staff will be able to advise you which will best suit the conditions in your garden. A good mix will have a main ingredient, such as bark, peat or sand, and be supplemented with the correct balance of the slow-release fertilisers and minerals that citrus do best in, including lime, iron, calcium and gypsum.

You need to make sure there is plenty of room for the roots, so select a large container (at least 90 centimetres across and of the same depth). Choose a container with gently sloping sides, not one that has a 'belly' or bulge halfway up, and plenty of large drainage holes. Cover the holes with shadecloth to prevent the potting mix from washing out, and to stop insects getting in.

Because of the weight of the planted pot, a small base fitted with castor wheels will make moving the container easier. If you have one, a greenhouse is an ideal winter location for container-grown citrus.

Pot your young citrus tree as you would any container-grown tree or plant. Firm it well into the mixture, making sure the tree is not planted too deeply — the surface roots should be just below the surface of the potting mix. Leave 7 centimetres from the mix to the top of the rim. You can place decorative pebbles over the surface of the mix to make it more attractive.

Container-grown citrus need a careful balance of watering, the same as any other container houseplant. You must maintain the balance between too much and too little. *Never* let the potting mix dry out, especially during summer when regular watering is essential. A drip-feed system to the root zone can be installed by using 4-millimetre micro-tubing connected to a water tap and regulated by an automatic water-controller.

Most watering problems are related to the potting mixture. There must be plenty of air space in the mixture after the water drains away. This is why a lightweight mixture of pumice and peat moss or granulated bark is so effective.

Citrus are heavy feeders and will exhaust the nutrients in a commercial potting mix within a few months. Long-term, slow-release fertilisers such as MagAmp or Osmocote can be combined with the potting mix. These last for up to 12 months. If they aren't in the original mix, start a feeding programme after five months. Supplementary feeding with a complete fertiliser such as Dig This Organic Liquid Fertiliser, a liquid blood and bone fertiliser made from waste from the fishing industry mixed with seaweed from Norway, can also be made once a week during summer and every two to three weeks during winter.

The feeding and watering requirements during winter will depend largely on the growing position selected. If the situation is warm and sunny, nutrient needs will remain fairly constant. Fertilise according to growth rate.

Citrus flowers alone may be reward enough for you and their fragrance will scent the whole house. If your trees are kept indoors or in a greenhouse all year and you want fruit from them, you will have to help pollinate some of the flowers. Use a small paintbrush or cotton bud to transfer the pollen from one flower to another. If there are many flowers on the tree, a good shake may be enough to spread the pollen.

The main problem experienced with indoor container-grown citrus is lack of humidity or moisture in the air. There is a clear relationship between humidity and plant stress indoors. At 24°C with a relative humidity (the amount of water vapour in the air) of 50%, plant stress is low. At 24°C but with only 20% humidity, plant stress is very high. The average humidity indoors during winter is usually less than 20%. This can create a great deal of stress when a citrus plant is brought indoors for the winter. Severe stress can cause lack of flowers, leaf drop, and some branches may even die back.

If the amount of water in the air stays the same, temperature will control the relative humidity. As the temperature drops, relative humidity increases. Also, lower temperatures slow the rate at which the plant grows, reducing its need for food and water. Try to have night temperatures about 10°C cooler than day temperature. A significant day-to-night difference will also promote a deeper orange colour in fruit.

You can increase humidity by placing the growing container on a large tray filled with pebbles and water. The pebbles keep the container above the water level. As the water around the pebbles evaporates, the humidity will increase significantly. Another way to increase humidity is to group plants together so that they give mutual protection from draughts and dry air.

Frequent misting is another excellent way to raise humidity. If the trees are in a greenhouse, you can set up a misting device operated by a timer.

Container-grown citrus will also better adapt to changes in humidity and growing conditions if gradually moved to their new position. This applies when moving trees outside as well as inside. Trees that have been inside should go to a shady position outside, or to a place that receives less than two hours of direct sunlight a day for a while, to avoid leaf burn.

Re-pot your containerised citrus every two years. Spring is the best time to do this, but do not delay if the leaves start turning yellow and dropping, or the tree shows other signs of poor health.

Tip the container on its side and, using a large knife, cut away any roots adhering to the sides. A jet of water will also help to loosen the tree and mix from the container. If the root ball is a mass of roots, remove the mix from the tangled outer roots and prune about 8 centimetres of growth from the outside and bottom of the root ball with a sharp knife. Place the tree back in the container with fresh potting mix. Water well and place in a warm, sheltered position until new growth appears on the tree.

11: Diseases and pests

MOST HOME GARDENERS will experience few problems with their citrus trees, especially if they follow a preventative spraying programme. Also, pest damage will be further minimised by growing healthy trees with proper water and feeding. Occasional problems may occur which warrant some control measures, but first consider how much damage is 'tolerable' before using pesticides. (See pages 91–94 for tips on spraying).

Fungus diseases

Alternaria brown spot

Alternaria brown spot attacks young fruit, leaves and twigs producing small brown-to-black spots surrounded by a yellow halo. The spots enlarge as leaves mature and some leaves may drop, or the entire shoot may die. Severe attack can cause young fruit to drop off and leave the remaining ones with spots, varying from small dots to large pockmarks on the skin.

The disease spreads by spores produced on infected leaves, both on the tree and on those that have fallen to the ground. Spores are air-borne and release is triggered by rainfall or by a sharp drop in relative humidity. The length of the wetting period required for infection to set in is about 8–10 hours when the temperature is around 20–29°C. Most of the infection probably occurs following rain, but dew is often sufficient to set it off. In the home garden, spraying with a copper-based fungicide, such as copper oxychloride, is the most practical means of controlling this disease. The number of fungicide applications needed varies greatly with the susceptibility of the variety and the severity of the infestation. In the worst cases, the first spray should be applied when new shoots are about 6–12 centimetres long to prevent build-up of the disease. The second application should be made at petal fall. Thereafter, applications may need to be as often as every 10 days to achieve good control on fruit and foliage.

Bark blotch

Bark blotch is a fungus disease that attacks the trunk and occasionally the branches, leaves and fruit. It can be difficult to detect as the first signs are the pale brown colour of infected bark and a slightly raised canker margin. On mature trees, the cankers slowly enlarge, usually growing along the trunk more rapidly than around it. On small branches the lesions will often girdle the branch, and the leaves will turn yellow and fall. The fungus will produce a firm brown rot in fruit.

These lesions should be removed with a sharp knife and the wound disinfected, then covered with pruning paste. Spraying with a copper fungicide, such as cupric hydroxide — by far the best one available — during wet winter weather will prevent this disease from recurring. Keep tall weeds away from citrus trees and keep mulching material at least 30 centimetres back from the trunk. This will reduce the chance of fungus diseases infecting the tree.

Citrus brown rot

Citrus brown rot, also known as collar rot when trunk and roots are infected, brown gummosis, foot rot and phytophthora leaf blight, is likely to attack lemons, oranges and New Zealand grapefruit. The infection appears on the roots and trunk to begin with, then the foliage turns yellow while the trunk exudes gum near ground level. Eventually the infected bark will dry out, shrink and crack. The brown rot kills the cambium, or growing layer, beneath the bark. Different rootstocks have varying degrees of susceptibility to brown rot. Trifoliata rootstock is very resistant to this fungus disease.

Leaf infection shows as translucent spots on leaf edges and tips. Some leaves rapidly turn brown, curl up, but remain attached to the branch for some time, while others will drop in a few days. Defoliation may be severe, particularly in the lower parts of the tree. The first signs of infection are usually found near ground level where the spores can be splashed up onto the tree, infecting fruit which then turns pale brown and rots.

Any soil-borne fungi attacking roots and trunks are difficult to control with chemicals. Control on leaves and fruit is much easier. First, cut off all low-growing branches to within a metre of the ground, then apply a copper spray in May, followed by another application in June. Selecting trees with a resistant rootstock is the best way to prevent infection of the roots.

Melanose

Melanose attacks the leaves, shoots and fruit of citrus trees. Soon after new leaves emerge, minute water-soaked spots appear, gradually becoming dark, raised areas with yellow halos. Leaves may become distorted and fall, while twigs become covered in scar tissue and die. Melanose spots also appear on the fruit as light brown, circular and sunken, later becoming dark brown to almost black with a waxy appearance. Melanose is most commonly found on lemons, oranges and New Zealand grapefruit.

The main infection time is during wet spring and autumn weather when the tree is growing vigorously. Prune out all infected and dead wood, then apply a copper fungicide spray in October (pre-blossom), late November (petal fall), and again in early January. The November spray is the most important.

Sooty mould

Sooty mould is a fungus which grows on the honey dew excreted by scale insects, or mealy bugs that bite into the bark to feed on the sap of the tree. The fungus itself does not harm plant cells or tissues but is a sign that insects have invaded the bark of the tree. Trees severely infected by sooty mould look as though they have been dusted with soot.

Check at your local garden centre for the most effective product (for example, a spraying oil such as Neem Oil) currently available for this disease.

Verrucosis

Verrucosis, also known as citrus scab and sour orange scab, infects the fruit, leaves and occasionally the shoots of citrus trees. Small, raised blisters appear on the fruit, eventually becoming light brown and corky so they stand out conspicuously on young green fruit. Small, semi-translucent spots also appear on leaves, gradually becoming well-defined corky pustules. Wherever lemons are grown they are liable to this severe infection, particularly on fruit. 'Clementine' and 'Kara' mandarins, limes, tangelos, and 'Wheeny' grapefruit are all susceptible in areas of high rainfall.

No fungicides capable of eradicating verrucosis lesions are known. Bordeaux or other copper-based sprays are the most effective control to stop new lesions forming on an infected tree. All infected plant matter will have to be removed as fungicides are capable only of restricting not eradicating verrucosis. Spray at petal fall in late November, January and again in May. Additional spraying in October and June may be required if verrucosis is severe.

Bacterial diseases

Citrus blast

Citrus blast affects grapefruit, sweet oranges and sour oranges. The leaf stalks blacken, the infection spreading to the base of the leaf. Infection usually follows physical damage and spreads from there. The diseased leaves fall, leaving cankers which take the appearance of small, red-brown scabs. Deep, circular, dark brown or black pits form on the fruit and some of these pits will join together to form extensive depressions.

Some measure of control can be obtained by providing shelter for the tree to reduce wind damage. Copper sprays applied after blossom fall in autumn will also provide control. Prune out and burn any infected shoots.

Viral diseases

Viral diseases have caused whole citrus industries to collapse in some countries. Over 60 million trees were lost in Brazil because there are no effective, practical treatments to cure citrus trees once they are infected by a virus.

Tristeza

Tristeza is the most serious of the viral diseases. The three main recognised disease symptoms are: wilting and often rapid death of infected trees (Tristeza decline); stem pitting, especially pronounced in grapefruit; stunting and chlorosis of seedlings of sour orange, grapefruit, 'Eureka' lemon and citron.

The disease is widespread in New Zealand but does not cause a disease problem when a tolerant rootstock in the form of trifoliata is used. The disease kills any trees grown on sour orange or lime rootstock through starvation of the roots. The sieve tubes just below the bud union die, preventing the transfer of sugars necessary for root function. Because infected trees die quickly the disease is also called 'quick decline'.

Tristeza is transmitted by grafting and by the black citrus aphid which is widespread throughout New Zealand.

There is a wide range of other virus diseases present in other countries that have not reached New Zealand. Quarantine regulations are essential in preventing the introduction of new strains of virus carried on imported fruit or budwood. Resistant rootstocks are the main defence against viral diseases, while the development of new rootstocks will provide the solutions for many viral problems.

Scaly butt

Scaly butt is caused by exocortis viroid. In order to manage the spread of this disease HortResearch set up a budwood supply scheme in the early 1990s, which unfortunately collapsed due to insufficient funds being available to cover the cost of running the scheme.

Viroids are mechanically transferred, so if growers select from healthy looking trees and sterilise their equipment with bleach when moving from tree to tree, then viroids are not likely to be a problem.

Citrus exocortis viroid (CEV), however, is transmitted by aphids and is found throughout New Zealand. Rather like the common cold, there are many harmful, or virulent, strains of CEV, but some are less virulent and have no symptoms. Once a tree is infected with a virulent strain it will not get reinfected with any other strain. The trick is to be infected with a symptomless strain, which will then protect the tree from more virulent strains.

CEV is managed in New Zealand by planting on resistant rootstocks and selecting scion wood infected with symptomless strains of CEV. Issues arose when CEV-free 'Navelina' and 'Newhall' navel oranges were imported and released in New Zealand. As they did not have the natural protection of mild CEV strains, some have been infected with severe strains. It is now a matter of waiting to select budwood from trees that have been grown in New Zealand for long enough to be infected with CEV and that show no symptoms.

Storage rot

When free from fungus diseases, citrus fruit can be stored for considerably longer than stone fruit. New Zealand-grown lemons will last for up to four months in the right conditions, while oranges, mandarins and New Zealand grapefruit will last for up to two months.

There are nine types of rot known to occur under New Zealand conditions in stored citrus fruits. **Green** and **blue moulds** cause the most wastage, attacking all types of fruit. They can only penetrate the skin of damaged fruit so careful handling is the best form of prevention. Even the smallest break in the skin will allow fruit-rotting fungi to enter. Commercial growers often bathe the fruit in a mixture of water and Captan and then use a spray of a synthetic hormone. This prevents the buttons falling away from the stem-end of lemons, reducing the risk of stem rot.

Insect pests

Citrus fruits have relatively thick skins, so their eating quality is not usually affected by pests to the same degree as other fruits. Nevertheless, insect attack can spoil the appearance of the fruit and also cause damage to the tree if control measures are neglected.

Black aphids suck sap from the tree, congregating in numbers on new growth. They also spread virus diseases. However, they can be controlled by predatory insects. Ladybirds, the praying mantis, aphis-lions and hover flies are all important predators of the aphid, while German wasps provide a natural control for caterpillars and fly maggots.

Parasites are also an effective natural control for many pests but, unfortunately, like the pest predators, they are often killed by sprays intended to control the pests.

Commercial orchards are usually protected from aphids and other pests by a variety of insecticide sprays. Biological insecticides are also in use by some citrus growers. Bacteria that cause a disease in pest insects are cultured artificially. The bacteria are then dehydrated into a powder for mixing with water and spraying later. Water reactivates the bacteria which kill the insects.

Lemon-tree borer attacks all of the commercial citrus varieties. The beetle is a native to New Zealand and flies usually in the early morning and evening. It lays eggs in the leaf and stem junctions of the trees, as well as any cracks in the bark. The larvae bore into the wood causing small twigs to die. The resulting cluster of dead leaves is very noticeable in late summer. Larger larvae creating larger tunnels can cause major damage, as branches then break under wind pressure or heavy crops of fruit.

Prune and burn any branches showing frass (the mixture of sawdust and excrement pushed out by the burrowing larvae). Push a wire down the tunnels to kill the larvae then pump pyrethrum into the hole using an oil can. Don't use petrol or kerosene as this can damage the tree.

Leaf-roller caterpillar damages the leaves, fruit and buds of citrus and many other plants. All leaf-roller caterpillars have a similar life cycle and cause a similar pattern of damage. The leaves are tied together or to the sides of fruit with silken threads made by the caterpillar. This provides a secure shelter within which the caterpillars chew pieces from the developing leaves and fruit.

Spray with Neem Oil or use Derris Dust to remove these pests. Spray every two or three weeks from December to January and again from February to April. Leave a clear period of three weeks before harvesting the fruit.

Thrips are small winged insects with rasping and sucking mouthparts. They move about on the tree sucking sap and excreting dark globules of waste fluid onto the leaves.

Control of thrips by chemical sprays is relatively easy, provided thorough coverage of foliage is achieved. Spray the foliage thoroughly several times through summer with Neem Oil or mineral spraying oil. Also remove and burn any infested foliage during autumn. Thrips tend to breed more rapidly under dry, hot conditions, and by regular watering or sprinkling to maintain a high humidity, populations of thrips can be kept under control.

Scale insects are a major pest in many home gardens. Heavy infestations lead to poor, distorted growth and eventually branches or even whole plants may die. Fruit also becomes discoloured with the insects' excreta as it turns black with the fungus, sooty mould. Although these crawling insects have a waxy, white armoured back, they can be eradicated with all-purpose spraying oil. This oil covers them and prevents them from breathing. Normal insecticides mixed with water will not penetrate the waxy covering of scales.

When crawlers are active, Neem Oil will clean them up very quickly. Several sprayings at two-week intervals are usually required. Mealy bugs also have a waxy, white coating covering their bodies. An all-purpose spraying oil will kill these pests. Both mealy bugs and scale insects can also be controlled naturally by ladybirds and lacewings.

Codling moth larvae may attack oranges. Neem Oil sprayed onto the trees from early October through until March will control these pests. Commercial lures are also available, usually through a good horticultural or orchardists' supplier. The lure is a natural organic pheromone which can be used to attract the moths to land on a pad of sticky paper. The lures usually last for five weeks but you need to ensure that you have the correct lure for the particular pest you are wanting to get rid of.

Citrus mites are another pest that occasionally infests trees. They are almost too small to see but can cause bronzing of leaves and twigs as well as a reduction in fruit size. Mancozeb, commonly used by commercial growers, sprayed in summer when the fruit is pea-sized, will destroy the mites; home growers should use Neem spraying oil.

The **guava moth** is a recent arrival from Australia. It has become a pest of citrus in Northland and is slowly but surely spreading south. HortResearch scientists predict that the guava moth has the potential to spread throughout much of the North Island and possibly even into warm parts of the South Island.

The female moth lays its eggs on the outside of the fruit and the young larvae bore into the flesh. Eventually the fruit falls to the ground where the mature larva leaves the fruit and builds itself a cocoon from leaf litter and small sticks on the ground. After pupating for about 14 days the adult moth, which is white with black speckled markings and about 10 millimetres long, emerges. Several generations are possible each year as the range of fruit guava moth can feed on ensures a year-round supply: guavas and feijoas in autumn; citrus in autumn, winter and spring; loquats in spring; plums, peaches and nashi pears in spring and summer; and macadamia nuts from summer through to early winter.

Because the larvae bore into the fruit immediately after hatching they are very difficult to kill with insecticides. At present probably the most effective method of control is to rake up all fallen fruit and leaf litter from beneath affected trees and bury or burn it, to destroy pupating larvae. Do this regularly — don't leave fruit lying on the ground under the trees for more than a few days as the larvae can pupate and the adult moth fly off within 14 days or so of fruit falling.

You can also try using a pheromone trap (as for codling moth on page 86) but ask the supplier which lure to use, as these need to be specific to the insect you are trying to trap. The lure should also be replaced at regular intervals as they lose their effectiveness over time.

Alternative sprays

Apply all sprays with caution, following the instructions given on the product and the safety advice on pages 91–94. In an emergency or for other advice, call the New Zealand National Poisons Centre on 0800 764 766. It is open 24 hours, 7 days a week.

Copper Oxychloride

- For any fungal attack, such as melanose, verrucosis, brown rot and citrus blast.
- Active ingredient: copper oxychloride.
- Application: apply at the rate of 15 grams per 5 litres of water. Apply pre-blossom and after petal fall.
- Caution — do not mix with lime sulphur.

All Purpose 3-in-one Spray Sachets

- For all insects including leaf-roller, mealy bug, aphids, bronze beetle, woolly aphids, caterpillars and spittle bug.
- Active ingredients: chlorpyrifos, carbendazim, mancozeb.
- Application: mix 1 sachet in 5 litres of water. Spray every 12–14 days until no more insects are seen.
- Caution — do not spray if bees are working the flowers.

No Insects Carbaryl 80

- To control insects on fruit trees including green vegetable bug, potato tuber moth, thrips and caterpillars, and codling moth on pip fruit.
- Active ingredient: carbaryl in the form of a wettable powder.
- Application: 1 level capful holds 5 grams. Apply at the rate 30 grams per 20 litres of water.
- Caution — do not apply if bees are working as it is extremely toxic to them.

Super Spraying Oil

- To control scale insects, including San Jose scale, thrips, citrus red mite, aphids and European red mite.
- Active ingredient: mineral oil.

- Application: year-round application; has no smell; no withholding period.
- Caution — always wear rubber gloves.

Maldison

- A broad-spectrum insecticide which will control many insects.
- Active ingredient: Maldison
- Application: 1 level capful holds 5 grams. Mix 10 grams or 2 level capfuls into each 5 litres of water. Spray every 10–15 days until all insects are gone.
- Caution — do not use on crops that will be eaten. Do not mix with Bordeaux or lime sulphur. Can be safely used in combination with Captan, Thiram, Zineb, and copper oxychloride.

Citrus-tree Borer Spray Injector Aerosol

- To control citrus-tree borer.
- Active ingredient: Permethrin.
- Application: inject directly into the holes in the trunk and branches. Apply at any time of the year.
- Caution — do not spray on to leaves as it may cause slight burning.

Insecticidal Dust Puff Pack

- Excellent for eliminating wasps in citrus trees.
- Active ingredient: carbaryl
- Application: locate wasp nest and apply the dust at dusk or evening.
- Caution — do not inhale the dust; wear a dust mask.

Thiroprotect

- An animal repellent which provides long-term protection against rabbits, hares and possums attacking young and old citrus trees.
- Active ingredient: Thiram.
- Application: small plants and trees — 20–40 centimetres high

will require approximately 5–6 millilitres per tree; trees 1 metre and over will require up to 20 millilitres. Apply immediately after planting with a fine sprayer to all areas accessible to hares and rabbits and possums.
- Caution — allow 2–3 hours' drying time after rain before applying the repellant.

Spray Stick
- Excellent for control of spittle bug. It is designed especially to break down the protective coating of the spittle bug, in conjunction with a suitable insecticide.
- Application: this is a biodegradable wetting agent suitable for use, as an additive, with garden sprays. It mixes easily with both acid and alkaline sprays. Add the required quantity to the spray tank after pesticide is mixed and just before tank is full. Stir well.

No Bugs Super Concentrate
- For the control of white-tailed spider. We have a huge problem in New Zealand with this spider and they are often found in citrus trees. This formulation is great for controlling these spiders and I do recommend you to think about using it as they bite and leave a very nasty wound which can have rather nasty consequences.
- Active ingredient: Deltamethrin, a potent and UV stable synthetic pyrethroid insecticide. Unlike the active ingredients in most fly sprays (which are based on Permethrin, as it is a cheaper insecticidal alternative and breaks down in UV light rapidly), Deltamethrin is UV stable and will retain its residual insecticidal efficacy for up to four months even in full, direct sunlight. In general, insects are 1000 times more susceptible to Deltamethrin than Permethrin, and white-tailed spiders are particularly susceptible to it.
- Application: apply as coarse spray to run off. The product is designed for both interior and outdoor use.
- Caution — wear protective mask, gloves and overalls.

Safety when spraying

As a general rule, modern pesticides are less toxic to humans and more environmentally friendly than they used to be, but there are still some basic safety principles you should follow.

The first and most important step is to choose the right product. At the time of purchase, read the label and check the following:

- Is it the correct product for your problem? If in doubt, ask for some expert advice.
- Are there any restrictions on its use, e.g. is it safe to use on food crops?
- What environmental precautions might be needed?
- Do you need specific equipment to apply it, e.g. a sprayer?
- Do you need specific protective clothing?

Before applying the pesticide, read the label again to find out:

- what protective clothing you need to handle the pesticide;
- how and when to apply it;
- what it can or cannot be mixed with (compatibility);
- its withholding period — how long after spraying must you wait until you can eat the sprayed fruit?
- about any warnings, precautions and first-aid measures; and
- any special instructions.

Hints for successful spraying

Wear protective clothing

Many chemicals can be absorbed into the body through the skin. The risk is greatest in hot weather when you are likely to be sweating more. Always wear a long-sleeved shirt, long trousers or overalls and a hat of some sort.

For some chemicals you may need gloves, waterproof outerwear (PVC) and gumboots. The label on the product will give the details of what is needed. Clothing worn during spraying should be washed after use. Goggles or a face shield will protect the eyes.

Protect yourself from inhaling fumes and sprays

A simple gauze nose mask (dust mask) will only protect you from some dust and liquid particles but it will not protect you from gases and vapours.

Use a cartridge or canister-type respirator when
- there are toxic dusts or vapours about;
- spraying indoors or in a confined space; and/or
- the droplet size is very small.

Equipment

To avoid the risk of cross-contamination and damage to desirable plants, have two sprayers, one for weedkillers and another for pesticides, clearly marked to avoid confusion. After use, wash the sprayer and mixing equipment out thoroughly with lots of water, allowing the waste water to drain onto soil, not into a stormwater drain or sewer.

Mix chemicals carefully

Greater precautions are needed when mixing chemicals than when spraying as you are handling the concentrated material when mixing. Make sure your measurements are accurate and clean up spills promptly. Always mix chemicals outside in the open air.

Avoid spray drift

Choose calm, still weather for spraying. Early in the morning or in the evening are often good times. Make sure the spray gets on only the target plants — avoid overspray drifting onto nearby plants or neighbouring gardens. The hazards of spraying to both yourself and others increase dramatically on windy days.

Personal hygiene

Immediately after spraying wash your hands and face and, if necessary, have a shower.

Store chemicals carefully

Store all chemicals in a cupboard or shed that can be locked and is safely out of the reach of children or animals. Over 60% of all poisoning with pesticides involves children under five years of age. It is your responsibility to ensure that children cannot come into contact with the chemicals.

Always leave chemicals in their original containers, but if you have to move one to some other container, at least make sure the container is not one normally used for food and drink. Relabel the new container carefully.

Dispose of empty containers carefully

Empty containers must be disposed of carefully. With most home garden pesticides triple rinsing of waterproof containers, puncturing to prevent reuse, then disposing of them with the household rubbish is usually sufficient. Don't burn cardboard or paper containers — put them inside a plastic bag then into the household rubbish collection.

First aid

Always wash off any chemical splashed onto any part of the body immediately. Splashes of chemical into eyes must be dealt with by immediately washing with water and continuing to do so for 15 minutes. Contact lenses must be removed before washing. Always see a doctor afterwards.

If chemicals are inhaled or swallowed check the product label for first-aid advice; remember instructions vary. If poisoning symptoms appear, seek medical advice. The first signs are usually nausea and headaches. Take the packet or container with you to the doctor. For urgent information about poisoning ring the National Poisons Centre on 0800 764 766.

Environmental concerns

Some insecticides are toxic to bees. By applying insecticides in the early morning or late evening while bees are not foraging, you will substantially reduce the risk of harming them or contaminating their honey with pesticide residue.

Avoid contamination of streams, ponds, drains and sewage systems. Never pour unwanted spray down the drain. Ideally, only mix as much spray as you need, but if necessary dispose of any extra by pouring onto the soil in a little-used part of your garden.

Spraying citrus trees

Choose a fine, windless day, preferably slightly cloudy without a hint of rain.

Start at the base of the tree. With the nozzle pointing upwards, thoroughly cover all parts of the leaves, branches and fruit. Then spray the upper surfaces. The nozzle should produce a fine mist-like spray to give efficient coverage.

Some form of spreader or sticker helps and many growers add Dig This Organic Liquid Fertiliser to the spray tank, which acts as a spreader or sticker as well as a foliar feed.

12: Monthly citrus calendar

August

Most citrus growth begins in August which is why this calendar begins on this month.

- Apply an all-purpose fertiliser around the trees at a rate of 500 grams for every year of the tree's age. Ensure that the soil is well watered first.
- Control emerging weeds but take care not to dig deeply as citrus are shallow rooting.
- All-purpose spraying oil may need to be applied to control mealy bugs. Mix 20 grams of spraying oil or Neem Oil with 10 litres of water and add this to 50 millilitres of spraying oil in 5 litres of water to make a general-purpose clean-up spray.
- Check for snails, which climb into the trees and strip bark from tender shoots.
- Check for drainage problems which can be rectified later in summer. If drainage is poor the trees will not be able to utilise the fertiliser provided to boost spring growth.
- Check the trees for lemon-tree borer.

September and October

- Rub out any new growth that may clutter up the centre of the tree.
- Spray for leaf-roller caterpillars if necessary.

- Check for ants feeding on honey dew in the trees. This is a sign of scale, mealy bug or aphids.
- Mulch around the trees to conserve water later in the season.
- Mix Dig This Organic Liquid Fertiliser with any sprays you are using to provide additional foliar fertilisers as a boost for the trees.

November

Flowering is now at its height and bees are busy pollinating the blossoms.
- Apply copper sprays as a protection against verrucosis and melanose. Copper oxychloride is the best. Mix it at a rate of 20 grams to 10 litres of water. Copper oxychloride will also help control citrus blast and citrus brown rot.

December

- Correct any drainage problems noted in August.
- Most fruit have set by now and many will drop early. This is a natural occurrence.
- Maintain a good supply of water to help swell the fruit.
- Spray with pyrethrum if leaf-roller caterpillars appear.
- Check for signs of any mineral deficiencies. Symptoms like pale green mottling between leaf veins may indicate the first signs of manganese deficiency. This can be corrected by spraying with manganese sulphate mixed at 10 grams per 10 litres of water.
- Another dressing of nitrogen applied as sulphate of ammonia or urea may be necessary if the spring was excessively wet.

January, February and March

- Spray regularly (every 14 to 21 days) for aphids and leaf-roller with Neem Oil. Add a fungicide to the sprays to restrict verrucosis (although you can't eradicate it completely), and also to deter snails.

- Continue watering to swell fruit size.
- Thrips may appear during very hot summers but will be controlled by insecticide sprays containing oil.
- Check for lemon-tree borer, remove any infected branches and burn them. Cover wounds with insecticide to deter the lemon-tree borer from laying eggs near newly cut surfaces.

April, May and June

- Control weeds before winter.
- If there is still plenty of new growth showing, apply an extra feeding of sulphate of potash at the rate of 28 grams per metre. This will help mature and harden any soft growth before winter frosts.
- Change your copper spray to cupric hydroxide in frost-prone areas. This will kill off bacteria that contribute to frost damage.
- Prepare new planting sites by deep digging and incorporating humus. Check that the site has good drainage.

July

This is the start of the new planting season in frost-free areas.

- Choose trees from the nursery with care, making sure they are healthy, not too old, and of a suitable variety.
- Remove any fruit from newly planted trees to encourage new growth.

13: Recipes using citrus

WHILE MANY OF US will have a trusted marmalade recipe made with grapefruit, oranges, lemons or limes, or a combination of our favourite citrus, we often forget, or are unaware, that these fruits can also be used in the kitchen in many other ways. In the following pages you will find many hints for food preparation using citrus, and a number of unusual recipes where the fruit plays a leading role. There are also some classic dishes which rely on citrus for their long-established appeal.

MARMALADE

The age-old art of preserving is a great way to enjoy citrus all year round. Marmalade must be the most popular of preserves: who can resist tangy marmalade thickly spread on hot buttered toast? It can even be served on ice cream, instead of a more traditional sweet sauce.

There are two quite different marmalades. Bitter marmalade can be made from pummelo, grapefruit, and Seville oranges. Sweet marmalade can be made from sweet oranges, mandarins, tangelos, lemons, limes and kumquats. When using the bitter varieties, the pith cooks to a clear, almost transparent colour, so these make marmalades with a very attractive appearance. Different varieties and types of citrus can be mixed together to give a wider range of tastes.

Hints for marmalade makers

When using citrus for cooking or marmalade, it is essential to thoroughly wash any dirt or marks off the fruit in warm water, and dry well before using. It is also best to use freshly picked, not-quite-ripe fruit, as they will have higher pectin levels than riper fruit, which will help the marmalade set better (see note on pectin below).

After it has cooled, the marmalade should be firm and clear, with the fruit evenly distributed throughout. If it is bottled while it is too hot, the fruit has a tendency to rise, so allow it to stand before bottling.

To avoid discoloration of the pith, use a sharp stainless steel knife; always use a wooden spoon for stirring, and use glass or stainless steel bowls for holding the sliced fruit before cooking. Never use aluminium containers as the acid of the fruit will react with the aluminium.

A frothy scum will form on the surface of the marmalade as it boils. This is normal and should be skimmed off carefully about 5 minutes before the end of the cooking time. As the marmalade is still quite hot when bottled, always warm the sterilised jars before filling. Cold containers and hot marmalade do not mix. They usually break and cuts or burns can result.

A brief note on pectin

Pectin, a soluble, gelatinous complex carbohydrate found in ripe fruits and some vegetables, is often used as a setting agent in jams and jellies. Fruits rich in pectin are peaches, apples, currants and plums. When fruits rich in pectin are cooked with the correct amount of sugar they will set like jellies. If the fruit is overripe, the pectin becomes pectic acid, which will not form a jelly when the sugar is added.

Setting test for marmalade

Drip the marmalade off a wooden spoon above the pan. If it forms large flakes instead of running off the spoon, the setting point has been reached. You can also place a little of the marmalade on a cold saucer and let it cool. If a skin forms on top which will wrinkle when pushed with a fingertip, it has reached setting point and is ready for bottling. To avoid overcooking, remember to remove the pan from the heat while testing for setting.

Problems that may occur
Shredded skin is tough? Insufficient boiling before sugar was added to the mixture. *Dark colour, sticky consistency and won't set?* Mixture was overboiled after the sugar was added.

Speedy Marmalade

Nearly all marmalade recipes recommend that sliced fruit should be soaked for between 24 and 48 hours, but a much faster method for home use is as follows. This method can be applied to any of the standard marmalade recipes included here.

Wash the fruit and cut in half. Slice the fruit finely or put through a kitchen whizz or blender.

Put the fruit and the water suggested in the recipe into a pan and bring slowly to the boil. Boil until zest (rind or skin of the fruit) is tender.

Stir in the sugar and bring back to the boil. Boil gently until setting point is reached. If required, add lemon juice to improve setting during this period. When mixture is cooked, remove from heat, stand for 6 minutes, then ladle into warmed jars and seal.

Grapefruit Marmalade

> **1.5 kg grapefuit**
> **4.5 litres water**
> **3.5 kg sugar**

Wash fruit carefully in warm water. Cut in half, slice fruit finely and place in a bowl. Cover with water and soak for 48 hours. Place fruit in a large saucepan, cover and bring slowly to the boil. Simmer for about 1 hour, or until the fruit is very soft.

Remove from the heat and add the sugar, then return to heat and stir until sugar is dissolved. Bring back to the boil and boil rapidly, testing for setting point from time to time. This may take anything from 10 minutes or longer, depending on the pectin levels of the fruit you use, so it's important to test for set at regular intervals. Be careful you don't overcook the marmalade past its setting point.

Remove from heat and allow to cool for 4 minutes so that the fruit will be evenly distributed. Pour into warmed jars and seal immediately.

Ginger Marmalade

6 oranges
2 litres water
juice of 2 lemons
250 g preserved ginger, chopped
2 teaspoons ground ginger
2 kg sugar

Wash the oranges in warm water. Use a sharp potato peeler to slice the zest (rind or skin of the fruit) into very fine slivers. Slice off any excess pith with a sharp knife.

Halve the oranges and squeeze the juice into a large bowl.

Chop up the orange flesh and add it to the bowl, together with the slivers of zest. Add the water and lemon juice and cover and leave to soak for 24 hours.

Place the fruit, zest, and all the fluid into a large saucepan. Bring to the boil and simmer for 1½ hours. Add preserved and ground ginger, and the sugar. Stir until sugar is dissolved, then bring mixture to the boil and boil rapidly for 10 minutes. Test for setting.

Remove from heat and allow to stand for 5 minutes. Stir briskly, then pour into warmed, sterilised jars and seal while hot.

Diabetic Marmalade

2 large lemons
1½ large oranges
300 ml water
25 g gelatine
5 saccharine tablets

Wash the fruit thoroughly in warm water. Peel off the zest (rind or skin of the fruit) thinly, being careful not to include any pith. Finely shred the zest and finely slice the fruit flesh. Place fruit, zest and water in a large saucepan and simmer for 2 hours.

After cooking, measure the liquid and make up to the original amount. Soak the gelatine in a little water, add to the marmalade and boil for 10 minutes. Just before removing mixture from the heat, stir in the saccharine. Do not boil the saccharine.

Pour into small, warmed jars, as this marmalade will not keep once opened.

Lime Marmalade

1.5 kg limes
3.5 litres water
2.8 kg sugar

Wash fruit in warm water. Peel off the zest (rind or skin of the fruit) thinly with a potato peeler, then cut the limes in half and squeeze out and retain the juice. Chop the remaining lime flesh finely.

Place the thinly sliced zest and chopped lime flesh into a saucepan with the water. Bring to the boil and simmer until the zest is tender.

Add the sugar, then bring back to the boil, stirring until the sugar is dissolved. Boil rapidly until setting point is reached.

Skim off the froth. Allow to cool for 10 minutes before pouring into warmed, sterilised jars. Seal immediately.

Lemon Marmalade

2 large lemons
5 cups boiling water
3 cups sugar

Cut the fruit into thin slices and put into a bowl. Cover with boiling water and leave to stand overnight.

Transfer fruit and liquid to a saucepan and cook gently until the zest (rind or skin of the fruit) is tender and the liquid reduced by half. Add the sugar and stir until dissolved. Then boil rapidly until setting point has been reached. Leave to stand for 10 minutes.

Stir gently; pour into warmed, sterilised jars and seal immediately.

Kumquat Marmalade

24 kumquats, rinsed and thinly sliced
2 oranges, rinsed and thinly sliced
8 cups water, or as needed
9 cups white sugar, or as needed
juice of 2 lemons

Combine the kumquats and oranges and measure them into a large saucepan. Add 3 cups of water per cup of fruit. Leave for 12 hours, or overnight.

Bring the fruit to a boil, reduce the heat, and simmer until the zest (rind or skin of the fruit) is very tender. Remove from heat, and measure the cooked fruit. Add 1 cup of sugar to the saucepan for every cup of fruit mixture. Mix in the lemon juice — about ¼ cup.

Return the fruit to the saucepan, and bring to a boil once again. Boil, stirring occasionally, until the setting point is reached, testing at regular intervals. Remove from heat, and skim any foam from the surface.

Transfer the mixture to warm sterilised jars, leaving a little headspace, and seal immediately. Refrigerate after opening.

GRAPEFRUIT

Grapefruit are rich in vitamins and mineral salts. The inner skin of the grapefruit is a source of quinine, so it's not surprising that the fruit sometimes has a bitter, though not necessarily unpleasant, taste.

• Thick-skinned fruits are usually the best value since they contain more juice than thin-skinned varieties. Even so, this thick skin is well worth candying and the resulting sweetmeat is much better than that made from most other citrus varieties.

• Grapefruit make excellent starters to a meal. Try halving the fruit and combining it with prawns or creamed spinach, served in the shells, or an eye-catching combination of compatible fruits such as kiwifruit, grapes, avocado, etc.

Grapefruit Margaritas

grapefruit slices, cut in quarters
salt
¾ cup grapefruit juice
170 g Tequila
56 g Triple sec*
2 cups cracked ice

Moisten rims of cocktail glasses with cut grapefruit. Swirl in mound of salt to coat edges. Combine grapefruit juice, Tequila, Triple sec and cracked ice in a blender. Blend until smooth. Pour into prepared glasses. Place a grapefruit quarter on each rim.

*Triple sec is the brand name for a strong, clear, orange-flavoured liqueur which has been distilled three times. Curaçao, Cointreau, and Grand Marnier are also triple secs.

Grapefruit and Orange Cheesecake

Crust:
2 cups crushed coconut biscuits
½ cup melted butter
2 teaspoons grated lemon zest

Filling:
2 grapefruit
3 oranges
3 eggs
⅔ cup sugar
⅛ teaspoon salt
½ cup orange juice
1¼ tablespoons gelatine
¼ cup water
500 g spreadable cream cheese
1¼ tablespoons lemon juice
2 teaspoons grated orange zest
1 teaspoon grated lemon zest
⅔ cup whipping cream

Crust: Mix together the crushed biscuits, melted butter, and lemon zest (rind or skin of the fruit). Firmly press onto the bottom of a 20-cm spring-form cake tin. Set aside.

Filling: Peel the grapefruit and the oranges and cut the segments into small pieces. Set aside.

Separate 2 of the eggs and combine the egg yolks, the remaining whole egg, sugar, salt and 1 tablespoon of the orange juice in the top of a double boiler. Place over simmering water and cook, stirring constantly, until the mixture thickens to form custard. Remove from the heat.

Soak the gelatine in the ¼ cup water for 5 minutes. Stir into the warm custard until dissolved.

Beat the cream cheese with remaining orange juice, lemon juice and zests until smooth. Beat into the custard. Fold in the grapefruit and orange pieces.

Lightly whip the cream and beat the egg whites until they form soft peaks. Fold the cream and egg whites into the cream cheese mixture. Pour into the prepared spring-form tin and chill for several hours or overnight.

Grapefruit, Romaine and Red Onion Salad

12 medium Romaine or other lettuce leaves, ribs removed
1 red onion, peeled, thinly sliced and halved
4 grapefruit, peeled and cut into segments

Lime Dressing:
mayonnaise, thinned with lime juice, to taste
salt, to taste
crushed red pepper

On each serving plate make a wreath of Romaine, or other lettuce leaves of your choice. Make inner circles with onions and grapefruit. Mix all dressing ingredients together well and pour over the salad.

Chicken and Fresh Grapefruit Stir-fry

1 grapefruit, peeled
1 x 230 g tin pineapple chunks
1 tablespoon cornflour
1 teaspoon soy sauce (preferrably low-salt)
2 skinless, boneless chicken breasts
1 medium clove garlic, minced
½ teaspoon vegetable oil
125 g snow peas, trimmed
2 spring onions, sliced diagonally

Peel and section grapefruit over bowl, reserving juice. Drain pineapple well, reserving juice. Combine juices and add enough water to equal 1 cup of liquid. Combine with the cornflour and soy sauce.

Rinse chicken breasts and pat dry. Remove any excess fat and cut into thin strips. In a large non-stick frying pan, sprayed with non-stick cooking spray, stir-fry the chicken with the garlic in oil over medium-high heat for 5 minutes, or until lightly brown.

Add the snow peas and cornflour mixture and cook, stirring until thickened. Add grapefruit, pineapple and spring onions and cook until evenly heated.

THE VERSATILE ORANGE

This fruit can be used in a hundred ways, on its own or with other fruits. Generally speaking, oranges, i.e. all sweet oranges, tangelos and mandarins, can be used in most recipes given for lemons, except that, as a rule, less sugar will be needed.

• Oranges are rich in vitamins, especially vitamin C. 'One fresh orange, 65 cm in diameter', says the United States Department of Agriculture, 'contains 66 mg of ascorbic acid, or a little more than the daily adult requirement of vitamin C.' The juice is rich in sugars and acids, one reason why it so delicious. Orange juice is used almost as much as the flesh of the orange. A glass is a good way to begin or end the day.

• Orange is also associated with meat in many dishes, either in the sauce, or sometimes, as with veal, cooked with the meat. Pork, ham, especially baked hams, and chicken go well with orange.

• Oranges are not often served with fish, yet orange slices make a good substitute when lemons are scarce, and orange butter is great with flounder.

• Orange zest (or rind), as well as the flesh and juice, is often used to flavour puddings, cakes, breads, biscuits and many desserts. It's a great idea to have a store of orange zest ready to hand to use when you need it. Pare the zest finely, let it dry naturally in the air after washing, and then chop or grind finely. For cakes and desserts this zest can be mixed with sugar — ½ cup zest to 2 cups sugar — and stored in airtight jars for later use.

• The essential oil of oranges is known as neroli. Scented orange-blossom water, manufactured commercially, is used in patisserie work and confectionery.

• Orange-flower sugar is used, as are all flavoured sugars, to make petits fours. Dry the flower petals by spreading them on paper towelling in a dry place. Add to them twice their weight in sugar. Pound the two together until they are well blended, and store in an airtight jar.

Orange and Watercress Salad

4 oranges
2 tablespoons olive oil
pinch sugar
salt
freshly ground pepper
a good bunch of watercress

Finely grate the zest (rind or skin of the fruit) of two of the oranges, avoiding any pith, then peel them with a sharp knife, removing all the pith. Run a sharp blade either side of the thin membranes, releasing whole segments one at a time. Let them drop into a bowl along with any juice that drips while you are performing this operation.

Make a dressing by combining the olive oil with 1 tablespoon of the saved juice, the finely grated zest, a pinch of sugar and a little salt and pepper. Toss the watercress with this dressing just before serving and scatter the segments of orange among the leaves.

Orange Braised Lamb Shanks

1 carrot, finely diced
1 medium onion, finely diced
2 celery sticks, finely diced
2–3 tablespoons olive oil
a few sprigs fresh thyme
2 bay leaves
2 garlic cloves, finely chopped
4 tablespoons sieved roasted tomatoes (or 1 tablespoon
    concentrated tomato purée)
½ bottle white wine
1 cup lamb stock (or water)
2 oranges, juice and finely grated zest (no pith)
1 lemon, juice and finely grated zest (no pith)
4 lamb shanks
salt and pepper
fresh parsley, chopped

In a suitable casserole, sweat the diced vegetables in some of the olive oil without browning, until tender. Add the thyme, bay leaves, garlic, tomato, wine and lamb stock or water, along with most of the orange and lemon zest (rind or skin of the fruit) and the juices (retain a few pinches of zest and a tablespoon or two of juice). Bring to the boil and lower to a gentle simmer.

Heat a little more olive oil in a separate frying pan and brown the lamb shanks on all sides, seasoning with a little salt and pepper as you go. Transfer to the casserole and cover with its lid. Cook in a pre-heated slow–moderate oven (about 150°C) until the meat is completely tender and coming off the bone, approximately 1½ –2 hours. Remove the shanks from the pan and keep warm while you finish the sauce.

Skim off some of the fat that is floating on the top of the liquid. Taste for seasoning and to assess its intensity. Boil to reduce if you think it needs it. Stir in the reserved juice to refresh the citrus flavour.

Serve one lamb shank on each warmed plate, with a generous amount of sauce spooned over. Sprinkle each shank with a little parsley and a pinch of the reserved zest. Accompany with mashed potatoes and some creamy beans, such as butterbeans.

Chocolate Orange Truffle Cake

500 g top-quality dark cooking chocolate
100 g very soft unsalted butter
zest of an orange
2 cups whipping cream
100 g icing sugar

To decorate (optional):
shavings of plain chocolate
crystallised orange zest (thin strips of blanched zest, dipped in beaten egg white then in caster sugar and left in a cool dry place to harden)

In a mixing basin, melt the chocolate gently over a pan of hot, but not boiling, water. When it is liquid, stir in the butter until it is very soft, but not quite melted. Add the orange zest (rind or skin of the fruit).

Leave the melted chocolate/butter mixture to cool (dip the basin briefly in cold water). As it cools, but before it sets, whip the cream and the icing sugar together until thick, but still soft and loose, then fold quickly but thoroughly into the chocolate and butter. Spread into a spring-form cake ring, sitting on a flat plate or cake board (on which you will serve the cake), levelling off the top with a spatula. Refrigerate for at least three hours to set firm.

Run a warm, damp cloth around the outside of the spring-form tin, then open the spring to release it, and carefully lift it off the plate. Decorate with shavings or gratings of chocolate, and the crystallised zest. Keep chilled until serving, then cut into slices and slide a warmed cake slice underneath to release them.

Orange Soufflé

This can be prepared in advance, leaving the whipping and mixing in of the egg whites until the last minute before placing the oranges in the oven.

2 large oranges
⅛ cup butter
⅛ cup sugar, possibly more
1 teaspoon cornflour
2 tablespoons cold water
1 tablespoon Grand Marnier liqueur
2 eggs, separated
1 tablespoon cream

Slice the tops from the oranges. See that they stand firm. (If they don't, remove a sliver from the base, taking care not to puncture the fruit.)

Scoop out all the orange flesh. (A soup spoon is ideal for this.) Dry and butter the insides. (Best done by melting a little butter and brushing it on.)

Blend or sieve the orange flesh and strain to collect all the juice. Boil this with the sugar to make a good syrup.

Mix the cornflour with the cold water and add to the syrup to make a sauce, stirring as it thickens over a low heat. Allow the sauce to cool.

Pour the Grand Marnier into the syrup, then beat in the egg yolks, then the cream. Taste and adjust the sweetening if necessary.

Beat the egg whites until stiff, and fold them into the orange mixture. Two-thirds fill the orange cases with the mixture, place in an ovenproof dish and bake in the oven at 190°C for 35 minutes. Serve at once.

TANGELO TREATS

Flounder with Tangelo and Ginger

4 flounder (or other flat fish, as available)
4 tangelos
4 tbsp butter
4 tbsp grated fresh ginger
4 tbsp chopped parsley

You can skin the flounders but it will have more flavour if you bake it with the skin on.

Place each flounder on one half of a sheet of greased aluminium foil, and fold up the sides to ensure the liquid does not leak.

Squeeze the juice from the tangelos, and pour equal amounts over each fish. (The slight sharpness of tangelos makes them ideal for cooking with fish. You might substitute the juice of three oranges here, provided it is mixed with the juice of a lemon to give a tart edge to the flavour.)

Break the butter into small pieces and place them over the surface of the flounder, then sprinkle an equal amount of grated ginger root over each fish. Fold over the other half of the aluminium foil, crimp the edges, and bake in a 190°C oven for 20 minutes.

If the flounder is baked with the skin on, remove the cooked skin with a sharp knife. Sprinkle with chopped parsley and serve.

Tangelo Muffins

1 large or 2 medium-sized tangelos
1 cup sugar
1 large egg
½ cup milk
100 g melted butter
1½ cups flour
1 tsp baking powder
1 tsp baking soda

Cut fruit into eighths. Weigh out 200 g. Put sugar and tangelo segments into a food processor and process until finely chopped. Add egg, milk and melted butter.

Measure dry ingredients into a bowl and tip in liquid mixture. Combine but don't overmix. Spoon into greased muffin tins and bake at 200°C for 12–14 minutes.

LEMONS — AN ESSENTIAL FOOD

Lemons are rich in vitamins, especially vitamin C, as well as mineral salts. Among its other qualities, the zest of a lemon is higly regarded for its medicinal properties.

- Lemon brings out the flavour in foods so you can be quite liberal with the juice. Hold a lemon under warm water for a few minutes and then roll it between your hands before squeezing it, to make the juices run more freely. When only a little lemon juice is required, rather than cut the fruit, pierce it with a fork and squeeze out only as much as you need. The skin will close later.
- Lemon juice contains citric acid. It is this acid which acts on other fruits and helps them to keep their colour, e.g. apples, bananas, pears and avocados. These all become discoloured when exposed to air after peeling, so squeeze lemon juice over them or soak them for a time in water with added lemon juice. Adding lemon juice to strawberry jam helps it keep a good colour and prevents the fruit from going grey.
- A slice of lemon or a little juice should be used when boiling kumara, Jerusalem artichokes, stems of Swiss chard and beetroot, or anything else that becomes discoloured when being cooked.
- Lemon juice can be added to sauces for fish or meat. Lemon juice makes a better vinaigrette and mayonnaise than vinegar. Whenever possible, in uncooked dishes, substitute lemon juice for vinegar, for it aids digestion and is better for one's health. Lemon juice has a good neutralising effect. Always serve lemon wedges with any kind of deep-fried food.
- Not only the juice but the zest or rind of lemons aids many

forms of cooking. Save the lemon zest after the juice has been squeezed from them and either dry and powder them or chop finely and freeze for future use. The white pith under the skin should be pulled away, because if the lemons are cooked with this a bitter taste may result.

• Rather than grate the zest, an operation which is wasteful because so much remains stuck to the grater, remove the peel finely by using a potato peeler. Cut this peel in strips and chop finely to dry or freeze. If it is to be used julienne, cut it into fine strips, blanch it, drain and dry before using in the dish.

• Lemons can be easily frozen. If you like to serve slices of lemon with drinks, wrap lemon slices into small packs with plastic wrap, then freeze them. Lemon juice can also be frozen using plastic ice-cube trays. When the juice is frozen, release the cubes and pack into bags.

Cold Lemon Soufflé

A delicious and quick-to-prepare dessert.

> **1 level teaspoon powdered gelatine**
> **1 tablespoon water**
> **3 eggs, separated**
> **juice of 1 lemon**
> **finely chopped zest of ½ lemon**
> **½ cup caster sugar**
> **whipped cream and lemon slices for decoration**

Dissolve the gelatine in the water, making sure that it has been properly softened.

Beat the egg yolks well and add the lemon juice, zest and sugar. Beat the mixture well. Add the gelatine liquid. Beat the egg whites until stiff and fold them into the mixture. Heap this into a serving dish and chill until set. Serve topped with whipped cream and decorated with lemon slices.

Whip cream with lemon juice and icing sugar to taste. Top at the last moment with almonds or pecan nuts browned in butter and sprinkled with brown sugar to give them a slight toffee finish.

Lemon Curd

450 g sugar
4 lemons, zest and juice
110 g butter
4 eggs, beaten

Put the sugar, zest (rind or skin of the fruit) and juice, butter and beaten eggs into a large basin and mix well. Place the bowl on top of a saucepan of simmering water. Cook carefully, stirring with a wooden spoon all the time, until thick. The curd should coat the back of the spoon. Pour into warm sterile jars, seal and keep in the refrigerator.

LIMES AND KUMQUATS

One can almost say that wherever lemons are used — juice, flesh or zest (rind or skin of the fruit) — limes can be substituted, and yet there are occasions when the subtle difference matters. The flavour of the lime is perhaps a little more pronounced, particularly in the zest.

Like the lemon, the lime is rich in vitamins, especially vitamin C. Varieties of limes differ in size and colour. The bright green fruits are usually thin-skinned and heavy. Yellow-skinned limes are not so acid as the green fruits, nor is the lime flavour so pronounced.

• Lime juice has a digestive or 'cooking' action on raw fish and is used in many tropical recipes for this purpose. Many fish caught in New Zealand waters can be prepared in the same way.

Raw Fish Salad

Dice any raw white fish into small cubes approximately 2 cm square and place in a deep glass dish. Extract enough lime (or lemon) juice from available fruit to cover the fish.

Mix in finely chopped onion, or spring onion, and green and red diced capsicum. Cover dish with plastic wrap and refrigerate until the fish is opaque (minimum 2 hours, preferably overnight). Coconut cream can be added if desired. (This option is an ingredient of many Pacific island marinated fish dishes.)

To serve, sprinkle with fresh tarragon if available.

Lime Julep

8 limes, squeezed
sugar to taste
½ cup brandy
soda water and ice
4 sprigs fresh mint

Pour lime juice, sugar and brandy over ice. Top up with soda water and garnish with mint.

Lime and Cumin Vinaigrette

2 tablespoons fresh lime juice
1 tablespoon fresh lemon juice
½ teaspoon ground cumin
½ teaspoon chilli powder
½ teaspoon salt
⅓ cup vegetable oil

In a bowl whisk together the lime juice, lemon juice, cumin, chilli powder and the salt. Add the oil in a steady stream, whisking all the time until the vinaigrette is thick and emulsified. Makes about ½ cup.

Sesame Lime Marinade

This tangy sauce is delicious with chicken or beef kebabs. It keeps the meat moist and flavours the vegetables.

1 tablespoon finely shredded lime zest
⅓ cup lime juice
3 tablespoons vegetable oil
1 tablespoon sesame oil
¼ teaspoon salt
2 tablespoons honey
1 teaspoon sesame seeds

Combine lime zest (rind or skin of the fruit), juice, oils and salt. Pour over chicken or beef and marinate for 1–4 hours. Prepare the meat for grill, reserving the marinade. Combine ¼ cup marinade, honey and sesame seeds. Baste meat while cooking and again before serving.

Lime Slice Biscuits

½ cup butter, softened
1 cup sugar
1 large egg
2 tablespoons fresh lime juice
grated zest of 1 lime
1⅓ cups all-purpose flour
1 teaspoon baking powder
½ teaspoon salt
sugar for sprinkling on biscuits

Beat together butter and sugar with an electric mixer until light and fluffy. Beat in egg, lime juice and grated zest (rind or skin of the fruit). Sift flour, baking powder and salt together, then add to egg mixture. Form dough into a 25-cm log and wrap in wax paper. Chill dough until firm, at least 4 hours.

Heat oven to 180°C. Line baking trays with baking paper sheets.
Remove wax paper from the dough and cut log into ½ cm-thick
rounds. Place the slices onto the prepare baking trays. Sprinkle tops
with sugar and bake for 10–12 minutes, or until pale golden. Remove
to a wire rack to cool.

Kumquat Chutney

Make up a batch of this spicy condiment and keep it on hand to liven
up your table. It goes well with chicken, pork, lamb and curry.

6 navel oranges
12 fresh kumquats or 250 g jar preserved kumquats
1 red capsicum, seeded and chopped
1 green capsicum seeded and chopped
1 onion, chopped
1 cup raisins
2 cups cider vinegar
2 cups packed brown sugar
2 cinnamon sticks
6 whole cloves
2 tablespoons finely chopped fresh ginger
½ teaspoon cayenne pepper, or to taste
salt and freshly ground pepper, to taste

Cut the unpeeled oranges into ½ cm slices, and cut the slices into
6 or 8 pieces. Cut the kumquats into ½ cm slices. Combine all the
ingredients in a large saucepan and bring to a boil over moderate
heat, stirring frequently. Reduce the heat and simmer uncovered for
1 hour, stirring occasionally. Ladle into clean jars and seal. Will keep
refrigerated for up to 4 weeks. Makes about 5 cups.

Index